Innovation Analytics
Tools for Competitive Advantage

Innovation Analytics
Tools for Competitive Advantage

Editors

Nachiappan Subramanian
University of Sussex, UK

S. G. Ponnambalam
Vellore Institute of Technology, India

Mukund Janardhanan
University of Leicester, UK

World Scientific

NEW JERSEY · LONDON · SINGAPORE · BEIJING · SHANGHAI · HONG KONG · TAIPEI · CHENNAI · TOKYO

Published by

World Scientific Publishing Europe Ltd.

57 Shelton Street, Covent Garden, London WC2H 9HE

Head office: 5 Toh Tuck Link, Singapore 596224

USA office: 27 Warren Street, Suite 401-402, Hackensack, NJ 07601

Library of Congress Cataloging-in-Publication Data

Names: Subramanian, Nachiappan, editor. | Ponnambalam, S. G. (Sivalinga Govinda), editor. |
    Janardhanan, Mukund, editor.
Title: Innovation analytics : tools for competitive advantage / editors,
    Nachiappan Subramanian, University of Sussex, UK, S.G. Ponnambalam,
    Vellore Institute of Technology, India, Mukund Janardhanan, University of Leicester, UK.
Description: New Jersey : World Scientific, [2023] | Includes bibliographical references and index.
Identifiers: LCCN 2022030887 | ISBN 9781800610002 (hardcover) |
    ISBN 9781786349989 (ebook) | ISBN 9781786349996 (ebook other)
Subjects: LCSH: Product management. | Product life cycle. |
    Technological innovations--Management.
Classification: LCC HF5415.15 .I556 2023 | DDC 658.5--dc23/eng/20220805
LC record available at https://lccn.loc.gov/2022030887

British Library Cataloguing-in-Publication Data
A catalogue record for this book is available from the British Library.

For any available supplementary material, please visit
https://www.worldscientific.com/worldscibooks/10.1142/Q0293#t=suppl

Desk Editors: Nimal Koliyat/Adam Binnie/Shi Ying Koe

Typeset by Stallion Press
Email: enquiries@stallionpress.com

Printed in Singapore

About the Editors

Nachiappan Subramanian (Nachi) has 25 years of experience, including 23 years in academia and 2 years in a consulting company. He is Professor of Operations and Logistics Management at the University of Sussex Business School. He is a Fellow at the Chartered Institute of Logistics and Transport (CILT), UK. Nachi has been recognized for his research and teaching excellence. To date, he has published more than 101 peer-reviewed refereed journal articles. Nachi has secured and successfully completed projects worth £1 million. His research areas are sustainable supply chains (environmental and social/humanitarian issues), risk and resilience, technology-enabled operations and marketing interface, and performance measurement. On the teaching front, Nachi is a Senior Fellow at the Higher Education Academy, UK. He has also written several books, book chapters, and cases, among which his case titled "Supply Chain Optimization at Madurai Aavin Milk Dairy" has been reported as the second largest selling Harvard case in India. Nachi has authored a book titled *Blockchain and Supply Chain Logistics: Evolutionary Case Studies* along with A. Chaudhri and Y. Kayıkcı, published by Palgrave McMillan. Nachi has edited a popular book titled *Supply Chain Management in the Big Data Era* published by IGI Global, Hershey, USA along with Hing Kai Chan and Muhammad Abdulrahman. His second book with IGI Global was titled *Industry 4.0 and Smart Manufacturing in Hyper Customized Era* edited along with S. G. Ponnambalam, M. K. Tiwari, and W. A. Wan Yusoff.

S. G. Ponnambalam has 39 years of experience, including 36 years in academia and 2 years in a manufacturing company as a Production Engineer. He is currently a Professor (Higher Academic Grade) at the School of Mechanical Engineering at Vellore Institute of Technology, Vellore, India. Earlier, he served as a Full Professor at the Faculty of Manufacturing and Mechatronic Engineering Technology at Universiti Malaysia Pahang and at the School of Engineering at Monash University Malaysia. He is a Fellow at the Institution of Mechanical Engineers (UK) and a Chartered Engineer registered with the Engineering Council, UK. He is also a Senior Member of IEEE, USA. His area of expertise includes manufacturing, supply chain management, optimization, and swarm robotics. He has supervised a number of doctoral students in the area of manufacturing automation and swarm robotics. He has organized many international conferences in the fields of automation, mechatronics, and manufacturing engineering. Ponnambalam has edited a book with IGI Global titled *Industry 4.0 and Smart Manufacturing in Hyper Customized Era* along with Nachiappan Subramanian, M. K. Tiwari, and W. A. Wan Yusoff. He has written over 300 articles published in various referred journals and refereed conferences, and has written chapters in edited books.

Mukund Janardhanan is a Lecturer in Engineering Management at the University of Leicester. He graduated with a Ph.D. degree in Manufacturing Optimization from Monash University in 2015. Prior to joining the University of Leicester, he was a Postdoctoral Fellow at Aalborg University, Denmark, and also worked in product development in an engineering design company. He has published over 40 articles in reputed journals and conferences. His research interests include Industry 4.0, manufacturing system optimization, and simulation. He has edited special issues in reputed international journals and has served as program committee member at reported international conferences. He has supervised several postgraduate student projects in collaboration with industries in the UK and Denmark. Mukund is a Fellow at the Higher Education Academy, UK, and a Chartered Engineer with the Institute of Mechanical Engineers.

Contents

Chapter 1

Introduction

Nachiappan Subramanian[*,‖]**, S. G. Ponnambalam**[†,**]**,**
Mukund Janardhanan[‡,††]
and Subramaniyan Mukund[§,¶,‡‡]

[*]*University of Sussex Business School, Brighton, UK*
[†]*Vellore Institute of Technology, Vellore, India*
[‡]*School of Engineering, University of Leicester, Leicester, UK*
[§]*Department of Industrial and Materials Science,*
Chalmers University of Technology, Gothenburg, Sweden
[¶]*Capgemini AB, Gothenburg, Sweden*
[‖]*n.subramanian@sussex.ac.uk*
[**]*ponnambalam.g@vit.ac.in*
[††]*mukund.janardhanan@leicester.ac.uk*
[‡‡]*mukund.subramaniyan@capgemini.com*

Innovation analytics (IA) is an emerging paradigm that integrates advances in the data engineering field, innovation field, and artificial intelligence field to support and manage the entire life cycle of a product and processes. In this chapter, we have identified several possibilities where analytics can help in innovation. First, we aim to explain using a few cases how analytics can help in innovating new products to the market specifically through collaborative engagement of designers and data. Second, we will explain the use of artificial intelligence (AI) techniques in the manufacturing context, which progresses at different levels, i.e., from process, function to function interaction, and factory-level innovations.

1. Introduction

Data are a rich resource that helps decision-makers overcome various assumptions and intuitions from product development to customer engagement. With respect to product development, design and data specialists must work together to enable companies to develop new products and business processes. As we are all aware, companies are heavily investing in data and design capabilities; however, those firms that could effectively use design and data capabilities to understand the challenges will be able to innovate substantially.

Innovation is the driver of economic growth; several empirical pieces of evidences from different countries suggest that innovation supports companies to develop and maintain competitive advantage. Innovation can be classified either as incremental or radical. Incremental innovation refers to stepwise improvements within an existing technological approach, whereas radical innovation refers to entirely new concepts or new ways to think about a product or a process.

Advancement in analytics and development of artificial intelligence (AI) techniques helps companies achieve incremental and radical innovation. Companies at this digital age strive for success and it is essential to adapt to the advancements in the field of analytics and utilize it for their innovation journey.

Innovation analytics (IA) is an emerging paradigm that integrates advances in the data field, innovation field, and artificial intelligence field to support and manage the entire life cycle of a product and process. Innovation analytics will become an integral part of the entire innovation life cycle that helps in making smart, agile decisions and also accelerates the growth of the businesses. Different types of analytics have been reported in the literature. The main four types are descriptive, diagnostic, predictive, and prescriptive. With the emergent techniques, we have an impression that the future of innovation happening in product and process developed will be by utilizing predictive and prescriptive analytics. By utilizing the analytical environment, companies will be able to drive innovation in their operational process and this will help them to remain competitive and successful in their domain market. Using analytics, one will be able to identify the features that are most and least popular and this will help the companies to develop products that perform better, are cost effective, and improve their processes. This chapter aims to understand the several opportunities and challenges involved in relation to innovation analytics.

Analytics methods such as data mining, predictive models, and simulation will help the companies to identify patterns and help them to drive radical innovation. The data and insights available will help in revolutionizing product development processes and methodologies. Data and analytics not only help in the innovation but also help in developing new data-based business models that will have unique value propositions. Utilizing analytics for innovation, businesses and companies will be able to make smart decisions, and the current growth in the digital tools and technologies will support them to innovate and grow.

IA as shown in Figure 1 plays a critical role in helping businesses and industries in making smart decisions. Companies are required to innovate based on ever-changing customer requirements and several opportunities provided by technological developments. It can be summarized that analytics is essential for innovation and will need to be an important asset for the organizations to compete in the competitive market. Companies should give a prominent place for analytics in their innovation strategies. Predictive analytics and simulation can help drive valuable innovation and companies must use internal and external data to drive innovation and reap rewards by using the data effectively.

In this chapter, we will discuss several key aspects involved in Innovation and analytics with respect to product development and processes improvement. We will also cover the current literature reported in the field and discuss how innovation analytics makes an impact on the businesses.

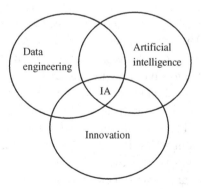

Figure 1.   Innovation analytics.

2. Innovation Overview

Innovation is mainly a process by which businesses/organizations develop new ideas, products, and processes. It is an essential for any business to be innovative to thrive in a global market. Different types of innovation that any business can pursue are related to individual products, internal processes/workflow, and even business models. Companies can try achieving all these together in an ever-changing market. Companies like Apple are a great example for an organization that has been successful in embracing innovation in its lifetime.

As per the literature, innovation has several definitions. In this chapter, we use Thompson's (1965) perspective and define innovation as generation, acceptance, and implementation of new ideas, processes, products, or services. Innovation can be classified in several ways (incremental, disruptive, architectural, and radical). Two major categories commonly described in the literature are incremental or radical innovation. Incremental innovation deals with doing what the company does but doing it better. In case of radical innovation, it needs to be something completely new and different (Tidd & Bessant, 2020). Incremental innovation refers to stepwise improvements within an existing technological approach, whereas radical innovation refers to entirely new concepts or new ways to think about a product or a process. Radical innovation has a higher risk and may not be successful all the time.

It is crucial for businesses to be innovative; otherwise, they are faced with the consequences of being driven out of the market. In the last decade, digital and technological advancements have forced several businesses to be ready for change and expansion. Businesses must be willing to adapt and change. Innovation helps businesses to grow, makes them relevant in a constantly changing world, and differentiates them from other competitors. The main concept of innovation is doing something different from others by developing, updating products or by optimizing the processes involved to save time, money, and other critical resources to achieve a competitive advantage in the market. Businesses must think creatively and embrace innovation in their eco-system. It has been extensively reported in the literature that innovation models for businesses can be classified in different ways. Based on the size, age, and the sector, the reasons to innovate will vary. However, before embarking in an innovation cycle, the organization will need to understand the different innovation models that are reported in the literature. One such innovation model can be based on revenue. To achieve this, the products, services offered will need to be carefully analyzed by benchmarking with other companies' pricing strategies. It is also to be noted that innovation need not be only the radical change to the system. Another model of innovation will be based on identifying the processes, products, or services where businesses can improve the profitability by developing new partnerships, implementing new technologies, or even outsourcing specific tasks.

3. Analytics

Analytics was derived from the Greek word *analytika*, which translates to "science of analysis". With respect to business, it is the analysis of large sets of business data, through analytical techniques such as statistics, mathematics, and computer software and applications. Analytics helps in capturing the status and changing within and outside an organization. It helps in providing real-time and predictive insights. Analytics can help organizations to develop new products or improve the existing processes. Data and analytics help in understanding new trends, developing inventive approaches, and achieving innovation in product development and processes. These techniques help in identifying new opportunities for businesses. Data obtained from different sources in the business can help in innovating and improving the business performance. Using the data obtained, one could improve and simplify the decision-making and create an innovative impact on the business models. Different types of analytics have been reported in the literature. The main four types are descriptive, diagnostic, predictive, and prescriptive. Using descriptive analytics, one will be able to answer the following fundamental question: What happened? In such analysis, using the events that occurred in the past, one tries to identify specific patterns within the data. This analytic methodology forms the foundation of other analytic techniques. In the case of diagnostic analysis, the main question that one attempts to answer will be as follows: Why did it happen? In this analysis, the organization must understand what happened to identify why it happened. Therefore, once the organization gets a good understanding of the descriptive insights, they will be able to apply diagnostics to get a deeper understanding of the situation. Predictive analytics is a form of analytics used by companies to plan a future product path and it tries to answer the following question: What is likely to happen? In this type of analysis, one will be required to use algorithms, data mining, and data interpretation to understand the market and trends. In predictive analytics, a collection of Artificial Intelligence tools such as machine learning, pattern recognition, sentiment analysis, and emotion recognition will be useful for the process of prediction. The most advanced level of analytics is prescriptive, and in this level, the question that one aims to answer will be the following: What should be done? Tools and techniques such as graph analysis, simulation, heuristics, and machine learning can be used to analyze the data.

Businesses should understand how to bring innovative ideas to life. One way of achieving this is by developing skills that are necessary within the businesses and one such focus area is Innovation Analytics where one must understand the different types of analytics available and how they can help businesses in their innovation journey. Although there are several reported research studies discussing analytics, there is a lack of clarity on how analytics and innovation can be linked, and this book will aim to provide clear insights into this. The central theme of this book is how companies can gain enhanced insights from analytics in their innovation journey.

4. Review of Research Studies

Analytics plays an important role in the innovation process. The innovation process involves lot of research and data analysis. It is critical to use analytics to help and support the innovation process. Companies must use internal and external data to help in their innovation journey. Using internal data, companies can make incremental advancements, which is considered the most common way of incremental innovation. External data that are available will help in generating new perspectives about the market and the performance of the company.

Due to rapid changes in the market, radical innovations are coming into the mainstream. Analytics methods such as data mining, predictive models, and simulation will help the companies to identify patterns and these can help industries to drive radical innovation. Data and analytics not only help in the innovation but also help in developing new data-based business models that improve the value proposition.

In this section, we performed a quick review to present a few emerging use cases addressed by the researchers on application of analytic tools to enhance innovation process. In particular, we realize that people still explore the impact of analytics on innovation, such as handling big data, tensions between governance and innovation, green innovation, and competitive advantage to the firms.

To sustain open innovation challenges in small and medium enterprises (SMEs) and multinational enterprises (MNEs), Del Vecchio *et al.* (2018) analyzed the impact of big data and analytics and offered a list of trends, opportunities, and challenges. In continuation of the above, Božič & Dimovski (2019) examined the relationship between the use of analytics and innovation ambidexterity. They tested a model using data collected from medium- and large-sized firms in Slovenia, applying partial least squares modeling. The results showed that the use of analytics is positively associated with successful balancing between explorative and exploitative innovation activities, which in turn enhances the firms' performance. In the context of UK SMEs, Liu *et al.* (2020) identified the barriers of utilizing cloud-based analytics for customer-driven design innovation in their study.

From the governance and innovation point of view, Mikalef *et al.* (2020) discussed the role of information governance in data analytics for driving innovation. They tested their model by collecting data from 175 IT and business managers. The finding states that data analytics would have a significant impact on both incremental and radical innovation capabilities of the businesses.

Use of data analytics to achieve competitive advantage is well explored. On the other hand, Zameer *et al.* (2020) explored the impact of analytics on green innovation and green competitive advantage through empirical evidence. Similarly, Duan *et al.* (2020) applied absorptive capacity theory to understand how UK firms recognize the value of data and how data are used to achieve competitive advantage. Their findings confirmed that analytics plays a significant role in sustaining innovation and helps in developing new products with the inclusion of environmental impact as well.

5. Analytics and Innovation

Analytics will help in bringing all the departments of the organization from sales to R&D to jointly collaborate in the innovation process and decision-making. Analytics will help in validating the creative process and providing necessary insights to the product innovation as discussed in the following subsections.

5.1. *Product innovation*

Analytics is vital to product development mainly with focus on product improvement. Without using analytics, the product development team will not be clear on what needs to be achieved and will not be able to meet customer requirements. Different metrics and understandings provided by analytics will help the product innovation teams to make informed decisions and improve the product functionality.

By using analytics in product innovation, businesses will be able to test, learn, and remodify the product design and launch process. This helps the decision-making process to be objective, reliable, and faster. Analytics can provide the team with real-time product performance measurement and helps in creating an accurate roadmap of how the product performed and what factors are to be considered in the future. Such analytical tools can also help businesses to understand their competitive advantage and help in proving a holistic view to the product team. Analytics will be able to give the most complete picture possible and this can be used to build the best product possible for the future. Advances in data management, cloud computing, and Big Data business will be able to enter a new age of product analytics. The data and insights available will help in revolutionizing product development processes and methodologies.

Applying analytics is a collaborative process that involves all the stakeholders, and analytics is applicable through every stage of the product development life cycle. Analytics must be embedded into every phase of the product life cycle development to gather data and insights to drive innovation from various perspectives. In regard to product development, managers and product developers have several critical questions to answer such as identifying factors and attributes that determine the future innovation in the product development. They are expected to understand internal and external factors that are essential for the success of a product. Identification of factors that will help in optimizing the product innovation, keeping customer requirement in focus, is also necessary. Mostly, companies have a limited set of tools and techniques and they heavily rely on the trial-and-error approach. However, the advancement of analytics can help in improving the product development process. Traditional approaches such as CAD, DFMA, FMEA, and Value Stream mapping are essential tools extensively used by organizations to improve the efficiency, eliminate waste, and optimize the product development process. However, as products are becoming complex and there is an abundance of data generated, these traditional tools are not being enhanced by the big sets of data and advances in the AI field to drive innovation in their product development process. Using analytics, organizations can improve their R&D process and also develop and identify new products and market segments. By applying analytics, organizations will be able to obtain clear customer insights and help in targeting the right audience for their products. It also helps them to price their products competitively.

In relation to innovation in product creation, using analytics, one can improve the performance by classifying the key product and customer attributes. Using this, one can model the relationship between the commercial success and the attributes to develop new products. In product testing, analytics can be used to model the relationship between the key performance drivers of the past products and the commercial success expected. Organizations can use similar concepts when planning and executing product launches, where key attributes of past success will be able to provide clear insights into what has to be done to achieve success in the future. The key aspect concerning product development is to collect data of the key attributes and develop the relationship based on the measure of success expected. To achieve this, a consistent, disciplined approach is essential.

Predictive analytics has been in focus in the last few years. The concept of predictive analytics can better help to understand the market, demand factors, and customer requirements. The scope of predictive analytics is enormous and can help achieve innovation in product development. Predictive analytics has not been extensively used in the field of innovation primarily because it is time consuming to develop predictive analytical models. However, it is very evident that predictive analytics will be crucial to achieve innovation targets in every industry. In predictive analytics, historical data are used to model the key trends and patterns. Using this model with the current data, one could predict what would happen next.

Application of prescriptive analytics helps to advance the innovation process in product development to the next level. Based on the predictive analytics, data organizations get an idea of what will likely happen in the future, so in that scenario, the organization will be able to plan what should be done. Using prescriptive analytics, one can develop various courses of action and outline what the potential implications would be for each. Prescriptive analytics applies artificial intelligence and machine learning, which includes algorithms and models that will help computers make decisions based on statistical data relationships and patterns. However, prescriptive analytics systems are not flawless, and they need to be monitored very carefully. There are concerns regarding the quality of data available and this can lead to wrong predictions. In the near future, several departments in the organization can use prescriptive analytics to drive innovation in the product development process.

5.2. *Process innovation*

In today's fiercely competitive global economy, manufacturing companies need to take quick and innovative actions, for example, swiftly measuring shop floor performance, rapidly identifying and trying new ideas, and scaling at speed as they grow. These require vast troves of factory data and tools that can analyze the data and provide critical insights into how the shop floor needs to evolve. That is why manufacturing companies are moving toward artificial intelligence-based (AI-based) solutions as a catalyst for their shop floor digital transformation and the key to making the right decisions to grow their business (Arinez *et al.*, 2020; Wang & Gao, 2020).

The manufacturing academic literature conceptualizes AI as a collection of tools that can extract insights from the manufacturing data and facilitate data-driven decision-making (Subramaniyan *et al.*, 2021). AI revolution is transforming the traditional manufacturing practices leading to increased productivity (Lee *et al.*, 2020). On the other hand, AI can help manufacturing companies innovate faster. For example, it can analyze interdependencies between humans, machines, robots, processes, and products more apparently by studying the data relationships. The new insights from these analyses can help shop floor engineers and managers to better understand the dynamics and make informed decisions.

Take the car manufacturing factory as an example. The factory will have several machines performing several operations (such as welding and painting) working together to manufacture a car. Consider this factory operation as a three-level hierarchical setup as described in Arinez *et al.* (2020). The bottom level, also called a process level, is where the tool interacts with the raw material to produce a car. At the middle is the machine level, where several components in a machine work together to execute a process. The top is the factory level, where several production resources, such as machines and robots, interact to establish a production flow.

At every level, engineers need to control numerous variables to produce a car. In addition, several variables at each level interact with one another, thereby increasing the complexity even more. Analyzing the variables and their interactions is necessary to identify the problems and design the solutions for productivity improvement. This is where AI can drive innovation. AI can explore all possible relations between the variables and hidden patterns that may not be readily observable by mining through the data. This leads to new insights and innovative solutions.

Let us look into how AI can drive innovation in eliminating throughput bottlenecks in a production system. Throughput bottlenecks are a high-frequency and high-impact factory-level shop floor problem (Subramaniyan *et al.*, 2016, 2021). Throughput bottlenecks are those machines or processes in the shop floor which constrain the shop floor throughput (Roser *et al.*, 2001). When engineers eliminate throughput bottlenecks, it is possible to get more throughput from the shop floor. In the real world, the engineers should identify the throughput bottlenecks almost daily and eliminate them to achieve the target throughput of the day. The traditional practice was to identify them through manual shop floor observations (Lee *et al.*, 2020; Subramaniyan *et al.*, 2021). This practice is time consuming and a manually intensive task. However, with the rise of digitalization and AI, identifying them has become an easy and less time-consuming task.

AI can create new knowledge about throughput bottlenecks on the shop floor and drive innovative elimination actions (Subramaniyan *et al.*, 2021). Consider a shop floor that has a serial production system of 10 machines. Every machine takes a set of states (e.g., producing, part-changing, blockage, and starvation) during the scheduled production time (Roser *et al.*, 2002). When engineers analyze these states and manipulate them (e.g., optimize the setup times), they can achieve higher throughput from the production system (Roser *et al.*, 2003). Consider that every machine assumed 10 distinct states during the production time. From a systems perspective, then there are 10^{10}, which is equivalent to 10 billion machine–state combinations. These massive combination sets are challenging to analyze manually. Usually, the shop floor engineers pick the throughput bottleneck machines in the production system based on their experience or simple static heuristics. This practice will not be effective because of the changing dynamics of the production system, for example, machines degrading with time and the introduction of new products and new machines changing the system dynamics. AI (e.g., deep neural networks) could help in these situations. AI can analyze every combination of the machine–state set, predict the combination that affects the system performance the most, and thereby identify the throughput bottlenecks. In this process, AI can reveal new information about the production system dynamics and throughput bottlenecks that is hard to capture by manual analysis. Engineers can use this information to design innovative solutions for eliminating bottlenecks and achieve the target throughput.

Analytics can play a key role in improving the project management process and can help in managing critical projects. Management of critical projects mainly deals with taking critical decisions and this can be supported by using analytics-based methods such as data mining and machine learning techniques. By analyzing the project data, one can make better decisions and also solve typical project-related problems. Using analytics, managers can predict early signs of deviations with respect to budget, cost, and time and take necessary corrective actions. Using analytics, one could estimate the progress of the work and also predict the possible completion time of the project.

Deeper and insightful analytics can improve the resource utilization and forecasting of cost and revenue. It is predicted that in the next decade, manual-based project management-related tasks will be taken over by analytics-based techniques. This does not mean that it will replace anyone's job; however, these techniques will help in making very informed decisions to improve the quality of the project execution with regard to time and cost.

Analytics will help in systematic quantitative analysis of data, and project managers will have the ability to optimize resource scheduling and allocation on projects. This will help them to propose the best possible schedule of resources with the available team. Using analytics, one could review the work and time-off schedules of all the people available at work and help in preparing weekly productivity reports. In short, analytics provides plenty of opportunities to optimize the project management implement process.

6. Summary of Chapters

We classify the 11 accepted chapters in this book as per the themes, namely, product and process innovation, artificial intelligence, and data engineering. This section offers a brief summary of chapters within each theme.

6.1. *Product and process innovation*

The first 4 chapters out of the 11 accepted discuss the application of emerging analytics for product and process innovation. The authors in *Chapter 2* have explored consumer product innovation and opportunities for data analytics. The focus of this chapter is to explain the possibilities for the use of AI-enabled tools and processes that can synthesize and mine social media data for innovation. In addition to the above, the study identifies practical barriers to using big data for innovation including in-house data analysts, siloed data sources, mismatch between product development and rapid change of consumer trends, and lack of integrated data mining capabilities.

The use of a system architecture computational tool to gather design data for developing an interdisciplinary product is explained in the *Chapter 3*. In particular, the chapter proposes a methodology based on function–behavior–state modeling and a computer-aided design system to customize the product at an early design stage.

Chapter 4 discusses business model innovation and potential application of innovation analytics. Use of social media data and the application of the same for a new business model targeting SMEs are explained in this chapter through an innovation process framework that emphasizes the need for capability development and multidisciplinary teams.

Smart manufacturing implementation and the role of product and process innovations are explored in *Chapter 5*. The chapter identifies 18 factors related to smart manufacturing implementation in the automotive sector that have a major influence on product and process innovations. The authors engage a VIKOR-based multi-criteria decision-making methodology to prioritize the factors.

6.2. *Artificial intelligence*

In terms of artificial intelligence, *Chapter 6* discusses the linkage between technological advancements and ubiquity. In particular, it seeks the potential application of artificial intelligence, non-routing algorithms, and visualization for innovation process. The chapter picks a hypothetical example of a start-up company that specializes in integrated circuit manufacturing to explain the innovative personalization process through customer data collection and analysis.

Chapter 7 narrates the analogy of human sensory organs with the industry Internet of Things and lists various enablers with respect to vision, sound, touch, smell, and taste/quality. In addition, the chapter reviews the popular machine learning algorithms such as unsupervised learning, supervised learning, deep learning, and reinforcement learning and their applications in product design, warehousing, and logistics and core manufacturing. The authors also suggest a few futuristic applications such as digital twins, edge computing, and technological aspects.

Chapter 8 reviews the potential application of artificial intelligence to analyze the food characteristics data generated through Internet of Things interconnected sensors to reduce food waste. The chapter also suggests various criteria to select the right AI tool for IoT implementation and its scalability to other sectors.

The role of machine learning and machine reasoning for intelligent transport systems is discussed in *Chapter 9*. Using case studies and surveys, the authors demonstrate the use of data generated through latest technologies such as video imaging and thermal imaging for better traffic management. The authors select Portland as a reference and explore the use of data analytics to manage traffic without any hassles.

6.3. *Data engineering*

Chapters 10–12 discuss the developments in data and they include the role of big data, evolution of time series data, fuzzy data, and textual data. A brief summary of each study is given as follows.

Chapter 10 explores the innovative solutions to reduce multiple corrections in the forecasting support systems. Specifically, the study discusses the process innovation in forecasting and inventory management. Through a survey, the study identifies the 11 most common drivers that necessitate the frequent adjustments made in the forecasting time series data and the use of exponential smoothing methods used in multiple industries.

Chapter 11 proposes a newer method to deal with fuzzy data in the selection of alternate designs in an open innovation process. The chapter reviews the use of multi-criteria decision-making for evaluating the appropriate design for mobile robot chassis, considering criteria such as novelty, manufacturing cost, assembly time, design complexity, and manufacturing feasibility.

Chapter 12 analyzes the forecasting method for innovative products. The authors identify the impact of electronic word-of-mouth criteria to predict the most preferred products by the customers. The authors explain Facebook's prophet forecasting model to identify the nonlinear trends on daily basis.

References

Arinez, J. F., Chang, Q., Gao, R. X., Xu, C., & Zhang, J. (2020). Artificial intelligence in advanced manufacturing: Current status and future outlook. *Journal of Manufacturing Science and Engineering*, 142(11), 1–16. https://doi.org/10.1115/1.4047855.

Božič, K. and Dimovski, V. (2019). Business intelligence and analytics use, innovation ambidexterity, and firm performance: A dynamic capabilities perspective. *The Journal of Strategic Information Systems*, 28, 101578.

Del Vecchio, P., Di Minin, A., Petruzzelli, A. M., Panniello, U., & Pirri, S. (2018). Big data for open innovation in SMEs and large corporations: Trends, opportunities, and challenges. *Creativity and Innovation Management*, 27, 6–22.

Duan, Y., Cao, G., & Edwards, J. S. (2020). Understanding the impact of business analytics on innovation. *European Journal of Operational Research*, 281, 673–686.

Liu, Y., Soroka, A., Han, L., Jian, J., & Tang, M. (2020). Cloud-based big data analytics for customer insight-driven design innovation in SMEs. *International Journal of Information Management*, 51, 102034.

Lee, J., Ni, J., Singh, J., Jiang, B., Azamfar, M., & Feng, J. (2020). Intelligent maintenance systems and predictive manufacturing. *Journal of Manufacturing Science and Engineering*, 142(11). https://doi.org/10.1115/1.4047856.

Mikalef, P., Boura, M., Lekakos, G., & Krogstie, J. (2020). The role of information governance in big data analytics driven innovation. *Information & Management*, 57, 103361.

Roser, C., Nakano, M., & Tanaka, M. (2001). A practical bottleneck detection method. In B. A. Peters, J. S. Smith, D. J. Medeiros, & M. W. Rohrer (eds.), *Proceedings of the 2001 Winter Simulation Conference* (pp. 949–953). IEEE. https://doi.org/10.1109/WSC.2001.977398.

Roser, C., Nakano, M., & Tanaka, M. (2003). Comparison of bottleneck detection methods for AGV systems. In S. Chick, S. P. J., D. Ferrin, & M. D. J. (eds.), *Proceedings of the 2003 Winter Simulation Conference* (pp. 556–564).

Roser, C., Nakano, M., & Tanaka, M. (2002). Shifting bottleneck detection. In E. Yucesan, C.-H. Chen, J. L. Snowdon, & J. M. Charnes (eds.), *Proceedings of the 2002 Winter Simulation Conference* (Vol. 2). https://doi.org/10.1109/WSC.2002.1166360.

Subramaniyan, M., Skoogh, A., Bokrantz, J., Sheikh, M. A., Thürer, M., & Chang, Q. (2021). Artificial intelligence for throughput bottleneck analysis — State-of-the-art and future directions. *Journal of Manufacturing Systems*, 60, 734–751. https://doi.org/10.1016/j.jmsy.2021.07.021.

Subramaniyan, M., Skoogh, A., Gopalakrishnan, M., Salomonsson, H., Hanna, A., & Lämkull, D. (2016). An algorithm for data-driven shifting bottleneck detection. *Cogent Engineering*, 3(1), 1–19. https://doi.org/10.1080/2331191 6.2016.1239516.

Thompson, V. A. (1965). Bureaucracy and innovation. *Administrative Science Quarterly*, 10(1), 1–20.

Tidd, J. & Bessant, J. R. (2020). *Managing Innovation: Integrating Technological, Market and Organizational Change*. Hoboken, NJ, USA: John Wiley & Sons.

Wang, P. & Gao, R. X. (2020). Transfer learning for enhanced machine fault diagnosis in manufacturing. *CIRP Annals*, 69(1), 413–416. https://doi.org/10.1016/j.cirp.2020.04.07.

Zameer, H., Wang, Y., Yasmeen, H., & Mubarak, S. (2020). Green innovation as a mediator in the impact of business analytics and environmental orientation on green competitive advantage. *Management Decision*, 60(2), 488–507. DOI: 10.1108/MD-01-2020-0065.

Part 1

Product and Process Innovation

Chapter 2

Consumer Product Innovation and the Opportunities for Data Analytics

Heather Burgess[*, §]**, Kripa Rajshekhar**[†,¶]
and Wlodek Zadrozny[‡,||]

Simpactful, Danville, CA, USA
†*Computer Science and Data Science, UNC Charlotte,*
Charlotte, NC, USA
‡*Metonymize, Deerfield, IL, USA*
§*hburgess@surprisinglyobvious.com*
¶*kripa@metonymize.com*
||*wzadrozn@uncc.edu*

The exponential increase in consumer behavior and product marketing data has not led to a corresponding increase in innovative output at leading Consumer Packaged Goods (CPG) companies. A number of practical barriers remain to the use of Big Data by innovation teams, including limited in-house data science talent, siloed data sources, mismatch in timescales between product development and rapidly changing consumer trends, and lack of integrated e-commerce and Web/Social insight mining capabilities.

On the other hand, many promising advances in machine learning, e.g., Natural Language Processing and Third-Wave AI, can help accelerate the convenient use of abundant consumer data for quicker and more efficient CPG innovation.

Using a combination of web surveys and face-to-face interviews, this chapter has identified practical concerns of leading innovation executives and their recommendations for technologists, e.g., long-term trend prediction and more direct links to sales/share metrics, as well as going beyond incremental performance marketing to more creative use cases.

Responding to practitioner needs, we offer examples of the most promising tools, areas of technology, and illustrative case studies. Starting with practical problems in mind, this chapter offers consumer business leaders a glimpse of the art of the possible, with easy-to-apply examples of how AI-enabled tools and processes can help synthesize and mine e-commerce/Web/Social data for smarter, faster, and more efficient growth and innovation.

1. Introduction

Innovation is on everyone's mind, as shown in Figure 1, even more than artificial intelligence, big data, and deep learning.[1] A similar picture emerges in Google Trends (not shown here).[2] Obviously, innovation is context and culture dependent. It can range from nation-scale innovation (see Kottak, 1990) to enterprise innovation (see Kopka *et al.*, 2020; Behrmann, 2019) to *consumer product innovation*, which is the focus of this chapter. While there is a great deal of qualitative research on innovation, mathematical treatment and computational framing of the topic are major research gaps (Kauffman, 2019; Loreto *et al.*, 2016).

Moreover, innovation should also be viewed as a process of adaptation (George & Lin, 2017), a common feature not only of the artificial world (Simon, 2019) but also of the natural world, and as such subject to investigations resulting in finding relevant "laws" (Lamsdell & Braddy, 2010). Thus, *innovation analytics* can be defined as the use of data to enable innovation in products and services, and as a tool to help in the mutual adaptation of the consumer and the product.

Notwithstanding this broader context, in the core of this chapter, by *innovation* we mean a product cycle ending with new product launch. We are focused particularly on advancement in learning techniques and analytics for

---

[1] https://books.google.com/ngrams/graph?content=innovation\%2Cartificial+intelligence\%2Cbig+data\%2Cmachine+learning&yearstart=1900&yearend=2019&corpus=26&smoothing=3.

[2] https://trends.google.com/trends/explore?date=2005-12-09\%202020-12-09&geo=US&q=innovation,artificial\%20intelligence,big\%20data,machine\%20learning.

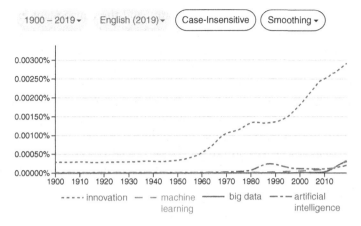

Figure 1. Sustained longterm interest in innovation is discernible Vs more recent hot-topics by comparing Google n-grams.

innovation, in the context of an exponential growth in behavioral and performance data via adoption of networked digital commerce, search and communication platforms (e.g., Amazon, Google, and Facebook).

To focus further, we will mostly talk about *consumer product innovation*. And, we explicitly exclude other types of innovation: service innovation, medicines, software and platforms innovation, etc. This includes offering insights and a real-world framework to leverage innovation analytics for competitive advantage, from new product development to operations, as well as enabling business models to experiment, test, and learn — faster, cheaper, and smarter — by building core competency in an emerging area of Innovation Analytics.

1.1. *Problem: The tempo of innovation*

Product introduction and adoption are faster than ever, which also means that product obsolescence is accelerating. Product cycle times are dramatically shorter than they were 20 years ago. Revenue contribution from new consumer products is steadily increasing, at faster rates (see, e.g., Driggs & Levin, 2020).

However, innovation is difficult and only about 30% of companies think they are good at it.[3]

---

[3] https://www.bcg.com/en-us/capabilities/innovation-strategy-delivery/overview. Last retrieved May 24, 2021.

Given the digitization of the economy, and the consequent explosion of data, it follows that many of the competitive problems companies have can be related to inference, speed of test–learn, adaptation, ideation, and data integration. Data mining of competitive, behavioral, and search data on the large and fast-growing platforms is an important source of practical innovation and testing capabilities (Leskovec *et al.*, 2020; Moe & Schweidel, 2012). Industry leaders have observed inefficiencies even in the more mature category of digital marketing (Pritchard, 2021). Therefore, a natural question arises as to the extent to which current data-driven techniques can improve the efficiency and speed innovation and what the impediments are.

1.2. *Data-driven models for marketing innovation*

Manipulating big data to create algorithms that optimize brand plans is not new. Google, Facebook, and Amazon spawned a 336B digital advertising industry built by leveraging 1st- and 3rd-party consumer behavioral data to form algorithm-based platforms that let advertisers address ads to potential customers based on increasingly predictive behavioral data (e.g., Erevelles *et al.*, 2016). In fact, according to a 2020 analysis by Statista, "over 51% of global media investments are now invested in internet media, which is powered by data-based algorithmic targeting."[4]

With Amazon Advertising, advertisers can target ads to customers based on viewing or purchase data down to a SKU level or a certain time window, and the Amazon Demand Side Platform (DSP) system will even enable marketers to auto-optimize campaigns through targeting, placement, or even the creative ads served to maximize the return on advertising investment against various campaign goals (O'Reilly & Stevens, 2018).

Similarly, trigger marketing or event-driven marketing represents a form of marketing that relies on the mining and modeling of important events within the customer life cycle to define not only the WHO to target but also WHAT content and WHEN to serve marketing efforts that drive awareness, consideration, or conversion, and will deliver the highest performance of a campaign (Dawson & Kim, 2010; Bughin *et al.*, 2010).

---

[4]https://www.statista.com/statistics/376260/global-ad-spend-distribution-by-medium/. Last retrieved on May 25, 2021.

In the past 5 years, advancements in data mining and modeling as well as the creation of systems to capture events and automate campaigns have set the stage for widespread adoption of these tools throughout the marketing funnel. LinkedIn estimates that trigger-based e-mail campaigns have five times the response rate of non-triggered push campaigns, and the effects can be seen across other forms of marketing from social and chatbots to phone calls.[5]

Marketers have adopted Big Data for downstream purposes (go to market plans) in large part because companies like Google, Amazon, Walmart, Facebook, Target, Shopify, and LinkedIn have intentionally created B2B monetization models which are built around the generation and access to this intuitively predictive data. They are selling a combined service package of not only the data and analytics but also insight generation, plug & play application (e.g., systems to buy media), and most importantly measurement. While the tools are not all perfect, marketeers are increasingly able to quantify the value of "performance marketing" (Vattikonda *et al.*, 2015), consistent with fund flows (Erevelles *et al.*, 2016).

While Big Data usage is common for go-to-market plans, our team sought to understand to what extent Big Data is being used to shape upstream disruptive innovation down to sustaining incremental innovation.

Switching to our focus on the CPG product innovation, we could ask the following questions: How do we find out what customers might be looking for? How do we present our products to the customers most likely to adopt them? What is the role of data and analytics in answering such questions? In this chapter, we attempt to answer such questions. To this end, we proceed as follows. The next section is intended to familiarize the reader with the problems faced by executives in the CPG space. We follow it, in Section 3, by examining results of our qualitative interviews with nine executives responsible for innovation at consumer product manufacturing companies at the forefront of using data for innovation in R&D and marketing.

Section 4 shows a few data mining case studies from Metonymize client work and those of other firms using AI to accelerate consumer product innovation. Our premise is that success in networked innovation is

---

[5]https://www.salesforce.com/products/marketing-cloud/best-practices/trigger-marketing/.

driven disproportionately by the identification and framing of important problem areas of immediate value to customers/users.

The discussion (Section 5) presents a view that the role of analytics in the innovation space is to help practitioners better represent and uncover affordances (Gaver, 1991) and latent semantic themes (Ahmad & Laroche, 2015) which might be used to more efficiently search through the space of possible innovations that meet consumer needs.

2. Major Themes of the Competitive Landscape

Product innovation is accelerating. Product adoption is faster than ever, which also means that product obsolescence is accelerating. In particular, we should note the following points:

- New product market failure rates remain stubbornly high (Schneider & Hall, 2011).
- Product cycle times are dramatically shorter than they were 20 years ago (Driggs & Levin, 2020).
- Revenue contribution from new consumer products is steadily increasing, at faster rates (Driggs & Levin, 2020; Services, 11 Nov 2020).
- Data mining of competitive, behavioral, and search data on the large and fast-growing platforms is an important source of practical innovation and testing capabilities (Leskovec *et al.*, 2020; Moe & Schweidel, 2012).

These points are connected to a collection of concrete problems and behaviors that need to be addressed to enable successful innovation. A typology of such problems and examples is shown in Table 1. The theme we will pursue in the next section (Section 3) is how they connect to analytics, and we will do it by analyzing results of the interviews with executives responsible for brand innovations.

We engaged nine practitioners who are actively leading innovation portfolios — including responsibilities across strategic "where to play" choice, in-house consumer and market understanding, and project scope. Our panelists serve across a variety of CPG categories industries from food and beverage and shave care to do-it-yourself DNA test kits and paper products.

Table 1. CPG brand innovators often try to solve common problems whether prompted by consumers, channel, or competitive needs. Data Analysis can support addressing most, if not all, of them.

Type	Description	Example
Purchase accessibility	Expand the number of places a product can be purchased	E-commerce enabling products
Usage accessibility	Expand the number of places a product can be used	On-the-go forms
Habit	Increase product purchase or usage frequency	Air care sprays with more displayable packaging to prompt usage
Purchase or usage appeal	Expand the number of people who can and/or want to buy or use the product	Children's or special audience variants/new products Improved claimable benefits
Value	Create an intuitively better benefit or a better value equation when factoring in ease, superiority against the core category drivers, proof of superiority, feedback loops, speed, and/or cost	Better flavors Ergonomics products concentrated benefits
Absence of a negative	Remove a real or perceived negative in the current product end-to-end experience	Natural products Hard seltzer (no hangover)
Obsolete an old category or create a new category	Innovate with the intent of addressing a new job to be done or finding a better/cheaper/easier way to accomplish an existing job to be done	2 in 1's hair color root touch-up sprays
Memorability	Make a product that cognitively en codes more easily or borrows familiarity to drive increased purchase and value	Trendy charcoal toothpaste or flavor variants; more distinctive packaging or branding
Business model enablers	Innovate to provide consumer benefit while also enabling new or existing business models	Smart products or experiences that provide consumer feedback or usage data; private label products; sourcing supply chain enabling innovation

While we interviewed a concentration of panelists from companies with annual sales of over 1.0B USD, we were surprised to learn that, qualitatively, 75% of companies are outsourcing data analytics. Key barriers to in-house analytics include access to data, modeling expertise, and competing resource needs to survive in the fragmented, omni-channel world where barriers to entry have evaporated for many companies. Today, our panelists report that they are sourcing data from a variety of places to guide their innovation strategies.

2.1. *Key drivers of big data perceptions and needs*

When we probed the extent to which big data is being used to guide innovation choices, almost all panelists, even from some of the largest CPG companies, felt they were "behind" the industry. They reported widespread use of algorithms to shape their downstream marketing plans, but were using predominantly purchase data, 3rd-party trend reports, and small data consumer research to determine innovation strategies and plans. This is shown in Figure 2.

- **Levels of Disruption:** The innovators we spoke to classified their innovation plans into four common buckets and leveraged this framework to help define their needs and where potential Big Data analyses could be impactful.
- **Commercial Innovation:** This involves the repositioning of existing products to drive top- or bottom-line growth with little to no invention required. This could include things like a new claim or new commercialization, a shift in artwork to enable cheaper printing of labels, etc.
- **Sustaining Innovation:** In this capacity, brands innovate product, benefit, component, or manufacturing lines using existing capabilities and existing markets. This could include things like new scents or flavors, more ergonomic packaging, new and improved formulas, and claims.
- **Breakthrough Innovation:** This type of innovation changes to an existing product, service, channel, or process that creates a significant impact on the business. For example, it could open up a new consumer category for the business, or change the way existing customers interact, buy, or use the brand.

Opportunities for Data-Driven Product Innovation: A Study

"What sources of information do you use to guide Innovation? (Mark all that apply)"

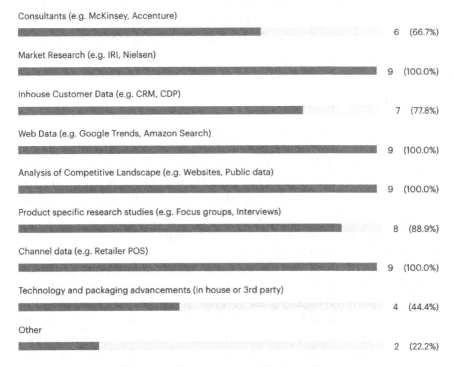

Figure 2. Data sources used for innovation.

- **Disruptive Innovation:** HBS professor Clayton Christensen developed the theory of disruptive innovation that changes the basis of competition. These new products or services enter at the bottom of the market or in new market footholds and over time move up and displace established market leaders, gaining mainstream appeal.
- **Time Horizon:** New products for CPG brands typically take 6 months to 5 years to bring to market. Most new product innovations require safety and regulatory work, formulation and packaging design, adaptations of manufacturing lines, and the creation of inventory to flow out into the omni-channel. There are infrequent exceptions that leverage

existing formulas, packaging, and manufacturing to enable speed while more disruptive inventions, such as connected products, those requiring multiple inventions, or those requiring new manufacturing capability, can require even longer lead times.

• **Data Source Fluency:** While our survey findings indicate respondents are using analytics to shape their strategies, few of the respondents we spoke to could volunteer mineable primary text data sources. They repeatedly mentioned importing analyses from suppliers such as IRI, Nielsen, and trend firms or, in some cases, deferring to internal market research teams.

2.2. *Innovator unmet needs*

We learned several key items when we asked practitioners which three key challenges they would prioritize for their innovation programs. Surprisingly, they did not seek to mine the data of competitors, to bring analytics in-house, or to integrate data from disparate sources such as Walmart and Amazon. Rather, their key challenges focused on (1) Defining unmet shopper needs, (2) Leveraging 3rd-wave analytics to predict opportunities and to forecast trends, and (3) to model the relative size of potential opportunities or ideas.

When we explored these unmet needs further, our respondents began to stitch together the challenge of managing both risk and the size of their innovation portfolio based on the level of disruption they were pursuing with their portfolio and the time horizon to bring the new products to market. For instance, the bigger the capital investment, the R&D investment, the breakthrough nature of category innovation, or the more competitive the category, the more value predictive analytics would offer.

3. Information-Driven Opportunities and Technologies

There is no one-to-one match between existing and emerging analytical technologies and the classes of problems identified in Table 1. For example, text analytics can be used to scout for new "Absence of a Negative" categories (by understanding social trends, extracting desired characteristics from negative reviews, etc.). They can also be used to improve "Memorability" and "Value". Therefore, we will look at five practical

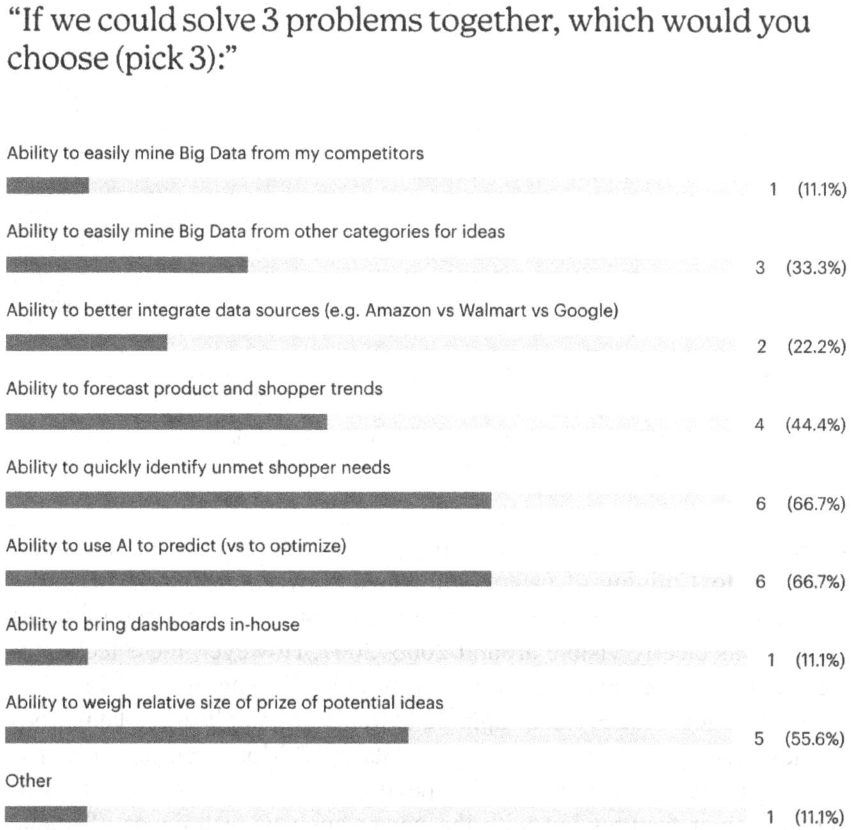

"If we could solve 3 problems together, which would you choose (pick 3):"

Ability to easily mine Big Data from my competitors

1   (11.1%)

Ability to easily mine Big Data from other categories for ideas

3   (33.3%)

Ability to better integrate data sources (e.g. Amazon vs Walmart vs Google)

2   (22.2%)

Ability to forecast product and shopper trends

4   (44.4%)

Ability to quickly identify unmet shopper needs

6   (66.7%)

Ability to use AI to predict (vs to optimize)

6   (66.7%)

Ability to bring dashboards in-house

1   (11.1%)

Ability to weigh relative size of prize of potential ideas

5   (55.6%)

Other

1   (11.1%)

Figure 3.   The use of AI to predict and the ability to identify unmet needs emerge as top problems in the CPG space.

examples of how information is being used to better understand opportunities for new products and better customer engagement. We will focus on the major issues shown earlier in Figure 3. We will proceed in the order from most general to most concrete.

3.1.   *Finding emerging technologies and products*

As shown in our previous work (Ankam *et al.*, 2013; Rajshekhar *et al.*, 2017), text analytics and machine learning enable one to find emerging technology trends. Figure 4 shows the growth and themes of patents in the

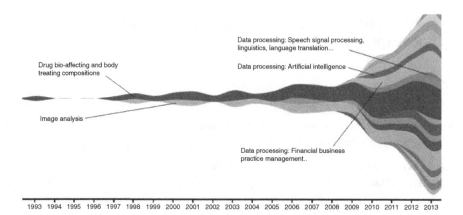

Figure 4.   This graphic shows the emergence of cognitive analytics and some of its major themes by text mining patent data from 1992 to 2013.

space of cognitive analytics. A similar picture emerged for smart phones. Based on text mining of patents published between 2001 and 2012, the 2013 article (Ankam *et al.*, 2013) showed that the emergence of smart phones was clearly visible around 2006–2007. However, these technologies are only infrequently used to understand emerging opportunities.

Similar techniques can be applied to trademark application data, product advertising, company registration data, etc.[6] Moreover, not only it is possible to extract descriptions of specific emerging categories of products and their features but with the newer analysis techniques of image segmentation (Minaee *et al.*, 2021) it is also possible to associate categories and features with both the specifics of their appearance in products and with the context of their use.

In addition to image analysis, related emerging technologies include video mining (Li *et al.*, 2019) and mining of mini-social networks (Leskovec *et al.*, 2020; He *et al.*, 2018), e.g., small networks on Pinterest, TikTok, or Facebook.[7]

3.2. *Using search and sales data*

Trend spotting and monitoring changes in product preferences across platforms are made possible by access to product search and sales data.

---

[6]https://developer.uspto.gov/open-analytics. Last retrieved May 23, 2021.
[7]https://www.theverge.com/2020/10/21/21526567/facebook-neighborhoods-feature-mini-social-network-sharing-test.

RISING.

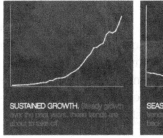

- waist trainer
- jogger pants
- palazzo pants
- tulle skirt
- midi skirt

- white lace dress
- high waisted bikini
- romper
- shift dress
- white jumpsuit

- neoprene swimsuit
- emoji shirt
- kale sweatshirt
- high neck bikini tops

DECLINING.

- one shoulder dresses
- peplum dress
- vintage clothing
- string bikini

- skinny jeans
- custom tshirts
- corset dresses

- normcore fashion
- 90s jeans
- scarf vest
- zoo jeans

Figure 5. Search and sales data can show typical rising and declining patterns. Understanding of the patterns can also be used in service of prediction.

In the following, we show, from a group of Metonymize clients, several visual examples of solving the problems identified in Figure 3.

Figure 5 shows typical patterns of growth and decline, and example products for each. The underlying technologies, using data integrated from various online sources, can be used for forecasting as well.

Figure 6 shows coffee product comparisons, and Figure 7 shows AI-assisted strategic segmentation. Both speak to the ability of weighing the relative prize size of repositioning the product (as well as to the data integration).

Analytics can be done in real time. Figure 8 shows our use of natural language processing to better understand the sources of happiness and unhappiness among sport fans. Note that real-time surveys, like the ones shown in the following, also enable real-time improvements, as well as limiting damage from negative experience, e.g., overflowing parking lots. Also, similar techniques can be applied to understand the degree to which new products satisfy customer needs.

Understanding the layout of the market and the marketing messages which accompany the products enable better placement and mapping of products to shopper preferences, perhaps based on the shopper's need states. This is shown in Figure 9 and speaks to the problem of identifying potentially unmet customer needs (a top concern in Figure 3).

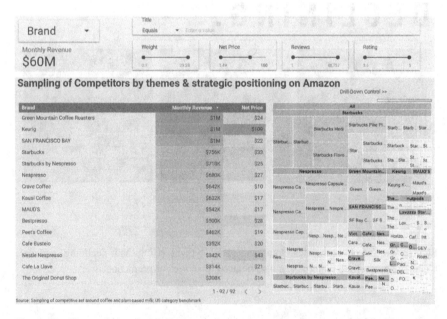

Figure 6.   Competitive advertising strategies can be analyzed and connected with product sales, giving a deeper understanding of the product market and potentially successful messaging.

Source: Sampling of competitive set around coffee and plant-based milk; US category benchmark.

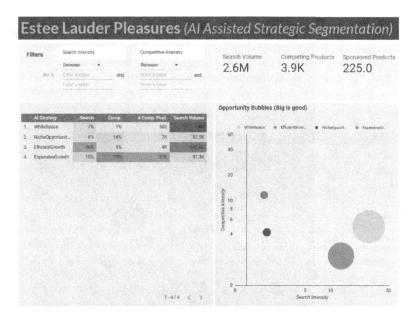

Figure 7.   AI-assisted strategic segmentation. Unsupervised clustering of search phrases and product descriptions from Amazon data, followed by analyst feedback to create strategic correspondences depicted on the left; the right-hand side shows that this helps identify opportunity areas where competitive positioning and responsiveness to shopper needs might be optimized.

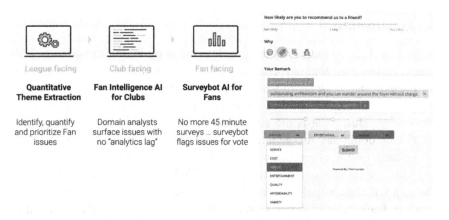

Figure 8.   Real-time surveys facilitate real-time reactions. Using human-in-the-loop model learning, a leading Sports League was able to dramatically improve listening to the "Voice of the Customer".

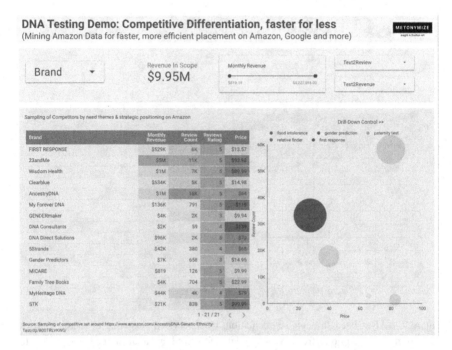

Figure 9.   E-commerce data mining in practice. Connecting classes of customer problems with market data (Metonymize, 2021).

Source: Sampling of competitive set around https://www.amazon.com/AncestryDNA-Genetic-Ethnicity-Test/dp/B00TRLVKW0/.

4. Discussion

In Section 1.2, we identified the problem faced by innovators in the CPG space: *How do we find out what customers might be looking for? How do we present our products to the customers most likely to adopt them?*

In Section 3, we showed examples of using analytics in service of innovation. But, what is the conceptual model behind all of these examples? One could say the general theme is of semantic mediation, that is, connecting customers with products and customer needs with new product ideas. Why is mediation necessary? Mediation is necessary because the explicit data which are available for analytics — searches, prices, product features, and other stored data — do not explicitly model the customer.

That is to say, they do not have information about the problem behind the search query, and neither do they have the utility functions of the customer, e.g., how much time are they willing to spend to find a better product, how much money are they willing to put in for a particular existing feature, and how much for a new product that solves the underlying problem. All of these — the not-yet-existing product, dollars/feature, and the underlying problem — are *latent*, and only emerge through a dissection of large amounts of data. Thus, the emerging model can be visualized as a *tripartite graph* where customers, analytic technologies, and producers are all linked. The role of semantic analysis is to mediate between these latent properties and the explicit attributes of products, i.e., product types, product features, and prices.

In many ways, we see the tools introduced above as a few illustrative examples of how we might expand consumer-centric versions of resources already available in the areas of patent data mining (Strumsky *et al.*, 2012; Goldschlag *et al.*, 2020; Bloom *et al.*, 2021). The idea of using graphs as models of business activity is not new; even the tripartite graphs have been used before, e.g., to model recommendation networks (Kim *et al.*, 2019).

The particular model we are proposing functions as a model of information-driven innovation, as it allows us to use the data as evidence for product gaps. Notice that this model only covers information-based aspects of innovation; missing are the physical aspects: matter and time. Bits and atoms are related: Being able to quickly produce a new or individualized version of a product would move the problem of CPG innovation from real-time analysis, shown in our examples, to real-time experimentation. Some of the enabling technologies are available, e.g., 3D printing, intelligent and reconfigurable production robots, and the IOT (the Internet of Things). Putting together such a system for CPG innovation would be highly non-trivial, but partial solutions exist, e.g., products (such as toothbrushes or air filters) that can reorder themselves.[8] So, on the small scale, paths connecting data and production exist.

5. Further Research

The mechanics and mathematics of evolution and innovation represent promising avenues for conceptual inspiration and further research. Entry points into this early but exciting area can be found in the work on

---

[8] https://www.wsj.com/articles/machines-that-shop-for-themselves-promise-to-save-time-and-money-11617807664.

"evolution without variation and selection" (Gabora & Steel, 2020), combinatorial network mechanics as indicated by the notion of the "adjacent possible" (Kauffman, 2019), and new probabilistic models for expanding spaces like the "urn model with triggering" (Loreto *et al.*, 2016).

One further area we are excited about is the concept of an affordance (Gibson, 1977), which we mentioned in the framing of the opportunities for innovation analytics in Section 1.2. This concept has been influential in Design thinking (Norman, 2004), including the area of human computer interaction (HCI) design (Gaver, 1991; Overhill, 2012). For instance (Gaver, 1991), using examples from Xerox PARC, e.g., Smalltalk, emphasized and extended the importance of Gibson's framework in the context of HCI with remarkable foresight from the twentieth century:

> Complex actions can be understood in terms of groups of *affordances* that are sequential in time or nested in space ... The notion of *affordances* is appealing in its direct approach towards the factors of perception and action that make interfaces easy to learn and use ... it allows us to focus not on technologies or users alone, but on the fundamental interactions between the two.

We see in this concept the potential for a type of universal representation of possible actions, extending the reach of innovation analytics into a more active and collaborative mode. One can assume that the Gibsonion notion of affordances can be linked in a straightforward manner to practical actions or latent semantic models as described in Section 4. Along with our emphasis on representing the consumer/user's intention unambiguously (as covered by some of the existing work (2, 4) and research in the field including (Ahmad & Laroche, 2015; Leskovec *et al.*, 2020; Rajshekhar *et al.*, 2017; Shalaby *et al.*, 2016; Zadrozny, 2006), we see *affordance orientation* as an exciting area for further research. This extends beyond innovation analytics to enable more active, intentional, and online-inventive collaboration of diverse types of humans and machines, furthering in some small way the cybernetic visions of individuals (Wiener, 1988; Bush *et al.*, 1945; Licklider, 1963).

6. Conclusion

In this chapter, we discussed the use of innovation analytics in product evolution and marketing, focusing on the consumer sectors. We showed how Big Data analytics tools can drive practical innovation by allowing companies to better respond to customer needs, especially when focusing on introducing new products. We also indicate areas of future research, building on the notion of affordances, while also leveraging the work being done in network models and the mathematical treatment of innovation. The interviews with innovation executives addressed the process of making analytics operational. We showed how integration of data across silos can show trends and quantify responses. We also proposed a novel tripartite graph model in which analytics drives practical innovation by allowing companies to better respond to customer needs, especially when focusing on introducing new products. In summary, we showed that data analytics allows companies to innovate and learn by being smarter, faster, and more cost efficient.

References

Ahmad, S. N. & Laroche, M. (2015). How do expressed emotions affect the helpfulness of a product review? Evidence from reviews using latent semantic analysis. *International Journal of Electronic Commerce*, 20(1), 76–111.

Ankam, S., Dou, W., Strumsky, D., Wang, D. X., Rabinowitz, T., & Zadrozny, W. (2013). Exploring emerging technologies using patent data and patent classification. In *Proceedings of IEEE VIS Workshop Interactive Visualization Text Analytics*.

Behrmann, K. (2019). Bursting with new products, there's never been a better time for breakthrough innovation. https://nielseniq.com/global/en/insights/analysis/2019/bursting-with-new-products-theres-never-been-a-better-time-for-breakthrough-innovation/.

Bloom, N., Hassan, T. A., Kalyani, A., Lerner, J., & Tahoun, A. (2021). *The Diffusion of Disruptive Technologies*. Technical report, National Bureau of Economic Research.

Bughin, J., Doogan, J., & Vetvik, O. J. (2010). A new way to measure word-of-mouth marketing. *McKinsey Quarterly*, 2(1), 113–116.

Bush, V. *et al.* (1945). As we may think. *The Atlantic Monthly*, 176(1), 101–108.

Dawson, S. & Kim, M. (2010). Cues on apparel web sites that trigger impulse purchases. *Journal of Fashion Marketing and Management: An International Journal*, 14(2), 230–246.

Driggs, J. & Levin, L. (2020). Innovation Before the "New Normal". https://www.iriworldwide.com/IRI/media/Library/IRI-New-Product-Pacesetters-Report-20.pdf.

Erevelles, S., Fukawa, N., & Swayne, L. (2016). Big Data consumer analytics and the transformation of marketing. *Journal of Business Research*, 69(2), 897–904.

Gabora, L. & Steel, M. (2020). Evolution without variation and selection. *bioRxiv*.

Gaver, W. W. (1991). Technology affordances. In *Proceedings of the SIGCHI Conference on Human Factors in Computing Systems* (pp. 79–84).

George, G. & Lin, Y. (2017). Analytics, innovation, and organizational adaptation. *Innovation*, 19(1), 16–22. https://doi. org/10.1080/14479338.2016.125 2042.

Gibson, J. J. (1977). The theory of affordances. *Hilldale, USA*, 1(2), 67–82.

Goldschlag, N., Lybbert, T. J., & Zolas, N. J. (2020). Tracking the technological composition of industries with algorithmic patent concordances. *Economics of Innovation and New Technology*, 29(6), 582–602.

He, K., Li, Y., Soundarajan, S., & Hopcroft, J. E. (2018). Hidden community detection in social networks. *Information Sciences*, 425, 92–106.

Kauffman, S. (2019). Innovation and the evolution of the economic web. *Entropy*, 21(9), 864. https://www.mdpi.com/1099-4300/21/9/864/htm.

Kim, K.-M., *et al.* (2019). Tripartite heterogeneous graph propagation for large-scale social recommendation. arXiv preprint arXiv:1908.02569.

Kopka, U., Little, E., Moulton, J., Schmutzler, R., & Simon, P. (2020). What got us here won't get us there: A new model for the consumer goods industry. In *Perspectives on Retail and Consumer Goods*. New York, NY, USA: McKinsey.

Kottak, C. P. (1990). Culture and "economic development". *American Anthropologist*, 92(3), 723–731.

Lamsdell, J. C. & Braddy, S. J. (2010). Cope's Rule and Romer's theory: Patterns of diversity and gigantism in eurypterids and Palaeozoic vertebrates. *Biology Letters*, 6(2), 265–269.

Leskovec, J., Rajaraman, A., & Ullman, J. D. (2020). *Mining of Massive Data Sets*. Cambridge University Press.

Li, X., Shi, M., & Wang, X. S. (2019). Video mining: Measuring visual information using automatic methods. *International Journal of Research in Marketing*, 36(2), 216–231.

Licklider, J. (1963). Intergalactic computer network. In ARPA. http://imiller.utsc. utoronto.ca/pub2/licklider_intergalactic_1963.pdf.

Loreto, V., Servedio, V. D., Strogatz, S. H., & Tria, F. (2016). Dynamics on expanding spaces: Modeling the emergence of novelties. In *Creativity and Universality in Language* (pp. 59–83). Springer.

Metonymize (2021). Metonymize marketing.

Minaee, S., Boykov, Y. Y., Porikli, F., Plaza, A. J., Kehtarnavaz, N., & Terzopoulos, D. (2021). Image segmentation using deep learning: A survey. *IEEE Transactions on Pattern Analysis and Machine Intelligence.*

Moe, W. W. & Schweidel, D. A. (2012). Online product opinions: Incidence, evaluation, and evolution. *Marketing Science*, 31(3), 372–386.

Norman, D. (2004). Affordances and design. Unpublished article. http://www.jnd.org/dn.mss/affordances-and-design.html.

Overhill, H. (2012). JJ Gibson and Marshall McLuhan: A survey of terminology and a proposed extension of the theory of affordances. *Proceedings of the American Society for Information Science and Technology*, 49(1), 1–4.

O'Reilly, L. & Stevens, L. (2018). Amazon, With Little Fanfare, Emerges as an Advertising Giant. *The Wall Street Journal.* Appeared in the November 27, 2018, print edition as *Amazon Emerges as Advertising Giant.* https://www.wsj.com/articles/amazon-with-little-fanfare-emerges-as-an-advertising-giant-1543248561.

Pritchard, M. (2021). Commentary: "Half my digital advertising is wasted...". *Journal of Marketing*, 85(1), 26–29.

Rajshekhar, K., Zadrozny, W., & Garapati, S. S. (2017). Analytics of patent case rulings: Empirical evaluation of models for legal relevance. In *Proceedings of the 16th International Conference on Artificial Intelligence and Law (ICAIL 2017)*, London, UK.

Schneider, J. & Hall, J. (2011). Why most product launches fail. *Harvard Business Review.* https://hbr.org/2011/04/why-most-product-launches-fail,13.

Services, H. B. R. A. (November 11, 2020). Delivering customer-focused innovation.

Shalaby, W., Rajshekhar, K., & Zadrozny, W. (2016). A visual semantic framework for innovation analytics. In *Proceedings of the AAAI Conference on Artificial Intelligence*, Vol. 30, p. 1.

Simon, H. A. (2019). *The Sciences of the Artificial.* MIT Press. https://doi.org/10.7551/mitpress/12107.001.0001.

Strumsky, D., Lobo, J., & Van der Leeuw, S. (2012). Using patent technology codes to study technological change. *Economics of Innovation and New Technology*, 21(3), 267–286.

Vattikonda, B. C., Dave, V., Guha, S., & Snoeren, A. C. (2015). Empirical analysis of search advertising strategies. In *Proceedings of the 2015 Internet Measurement Conference* (pp. 79–91).

Wiener, N. (1988). *The Human Use of Human Beings: Cybernetics and Society.* Number 320. New York, NY: Da Capo Press.

Zadrozny, W. (2006). Leveraging the power of intangible assets. *MIT Sloan Management Review*, 48(1), 85.

Chapter 3

R&D in Product and Process Innovation — System Design of Multidisciplinary Products by Applying Mass Customization Approaches

Tufail Habib

Department of Industrial Engineering, UET,
Peshawar, Pakistan
tufailh@uetpeshawar.edu.pk

In this study, a method is proposed to support the process of translating customer requirements into system requirements and transforming system-level specifications into subsystems and components at the early design phase of multidisciplinary products. System architecting (SA) is effectively applied to perform the functional decomposition of the product and then map the product functions into a physical structure such as modules and components. From SA, product architecture with interfaces is developed from which the modular product platforms are identified. In this study, a system architecture approach is applied in a computational tool (SA-CAD) to develop design data about the multidisciplinary product, i.e., autonomous vacuum cleaning robot. The result shows that the designer identifies the features and attributes of the product. The data stored in the computational tool can be used to establish consistency of system descriptions at different levels of hierarchy. The results demonstrate that the system designer gains the capability to redesign and develop multiple system architectures to facilitate product platform development and product family modeling for next-generation products. With this capability, companies can bring agility into their design, offer customized solutions to customers, and develop their production lines based on common parts and variable parts. Furthermore, this approach supports the task of system decomposition, complexity management, interface development, and the identification of system properties at multiple levels in a multidisciplinary product.

1. Introduction

The global economy is driven by rapid innovation, competition to introduce innovative products and processes, shortened product life cycles, and increasing customer demands in terms of the performance, quality, and cost of the products. Product and process innovations make a significant contribution, as innovation is one of the drivers to be competitive in the business. The application of new technologies such as cyber physical systems (CPS), Internet of Things (IoT), Internet of Services (IoS), and big data analytics is also shaping the industry. These changes have a significant impact on the entire value chain and demand new business models and approaches. To address these challenges, industries must be able to respond quickly and with flexibility in order to be competitive in the global environment.

The application of emerging technologies in manufacturing is changing the production systems rapidly. It enables industries to quickly adjust their physical and organizational structures and facilities. With these technologies, manufacturing becomes faster and more responsive to customer requirements and changing global markets. Researchers and practitioners argued that for businesses to cope with challenges and future uncertainties, mass customization, among others, is one of the solutions which has evolved into a flexible, fast delivery, and cost-effective production and marketing strategy. The best way to achieve mass customization is to create modular components that can be configured into a wide range of end products and services.

In order to be competitive, new models and methods must be introduced by the industry. Especially, in multidisciplinary systems design, designers need to deal with complexity derived from the integration of subsystems with various engineering disciplines. In particular, while developing product architecture for next-generation systems, the present generation systems should be reviewed in terms of their functional overview as well as their module structure. The objectives of this study are to develop a method to address the integration and complexity issues and incorporate customization at the early design phase of these products. Therefore, in this study, a function–behavior–state (FBS) modeling is implemented in system architecting tool i.e., (SA-CAD) to achieve is used to achieve the task of system decomposition, interface development, and complexity management at multiple levels in a multidisciplinary product.

2. Literature Review

In the last two decades, product modularity and product families have been advanced as effective approaches to achieve the purpose of mass customization (MC) to address the challenges faced by the industry. Over the last few decade, industries with high volume and a market for customized goods have embraced the mass customization strategy (Welcher & Piller, 2012; Nielsen, 2014). The best way to achieve mass customization, according to Pine (1993), is to create modular components that can be configured into a wide range of end products and services.

In MC strategy, process flexibility is one of the important elements. Two essential components of process flexibility are (1) the use of modular product design combined with delayed differentiation of a product and (2) the use of a flexible manufacturing system (Berman, 2002; Colombo 2020).

When a product or process is modularized, the elements of its design are split up and assigned to modules according to some architecture or plan. Pine (1993) has argued that for businesses to cope with challenges and future uncertainty, mass customization is a solution to be agile and responsive to global competition. Modular product architectures facilitate mass customization by allowing a wide range of products to be configured and assembled (Mikkola, 2007). According to Frederickson (2005), modularity can occur at various levels such as in products (product architecture designs), processes (manufacturing processes), and logistics (supply chain configuration).

Since the components of modular product architectures have standardized interfaces, mass customization and related manufacturing strategies are possible (Mikkola, 2007; Steffen, 2013). According to two core principles of MC, product ranges should be built on the basis of modules, and configuration systems should be used to support the tasks involved in customer-oriented business related to the specification of customer-specific products (Hvam, 2008).

The quest for potential technological solutions led to the introduction of modularity at various stages of development (Liu Zhou, 2010). The early stage modularization process allows for more freedom in defining architectural content and enabling the function–component mapping relationship.

Function-based module concepts can be used to explore conceptual product design and gain an early insight into common and specific functionality (Stone & Wood, 2000; Stone *et al.*, 2000; Dahmus *et al.*, 2001; Habib, 2020). Physical modularization creates modular product architecture by rearranging physical components into modules, and it is used to develop modular product platforms (Martin & Ishii, 2002; Hsiao & Liu, 2005).

2.1. *System design*

Originating from the field of system theory, a system is a mental concept and thus an abstraction used to explain and model a particular area of interest. A structure, in the context of technological structures, is concrete, complex, and made up of components. The structure of the system is defined by the relationships that exist between the elements. A system can be part of a larger system and can be decomposed into subsystems (Pahl & Beitz, 2008; Komoto, 2012).

As multidisciplinary systems such as mechatronic systems grow in size, system-level design and development become increasingly important, particularly in supporting complexity management, conceptual design, and domain integration to achieve desired results (Tomiyama, 2007). Some of the benefits of using system design approaches are:

- Identify interfaces between domains;
- Coordinate hardware and software;
- Increase the effectiveness or reuse of the design.

Via hierarchical system decomposition, system architecture (part of system design) of multidisciplinary systems describes subsystems and their interfaces. According to the V-model of product creation from a systems engineering perspective, conceptual design is also known as system architecture and is part of the system decomposition process (VDI 2206, 2004; Dieterle, 2005). Various authors (Pahl & Beitz, 2008; Chmarra, 2008; Hehenberger, 2009; Habib, 2014) have mentioned conceptual design. Since the product's overall functions, essential sub-functions, and their interactions are determined during this process of design, it plays an important role in system design. Principle solutions, as well as a system of realizable modules and their interactions, are determined in the conceptual design to achieve the successful design of systems. In addition, it is also necessary to perform these tasks with computational support.

One of the challenges in system architecture is how to depict mechatronic systems in a single model that is independent from all domains. According to Burr (1990), function modeling can be used across mechanical, electronic, and software disciplines, allowing for the use of function modeling approaches across mechatronic domains. Another issue in the early stage of mechatronic product design is how to effectively derive the parameters of a product and its subsystems on the basis of abstract descriptions of products, such as function requirements.

However, in order to complete the design tasks, the system's functional model should be developed concurrently with consideration of the actual physical environment or decomposition from function to form at multiple levels. In addition, to aid system decomposition and configuration tasks in mechatronics, the use of system design support tools is necessary (Komoto, 2010).

The major challenge facing the researchers studying conceptual design is to develop modeling schemes supporting the initial design processes. However, the modeling schemes should be independent of specific disciplines. One of the problems while modeling at the early design stages is not only to establish relationships between function and form based on the design concepts of a product and its subsystems but also how to address the consistency of system descriptions at different levels of hierarchy (Alvarez, 2010).

Furthermore, the creation of mechatronic products necessitates cross-domain integration, with particular attention paid to product dependencies and interactions between design activities. System designers must conduct system modeling not only by defining acceptable component level requirements but also by being able to transfer system information to communicate with domain experts. It is critical to address the problem of cross-domain collaboration and communication among design engineers in different domains.

Therefore, it has been recognized that in order to effectively utilize system design tasks in complex mechatronic systems, methods and tools are crucial. To meet these requirements, a new method for enabling the task of system decomposition, interface management, and the identification of system properties is proposed. The system architecture tasks must be supported by a computational tool to perform the above tasks.

3. System Architecture Using Function–Behavior–State Modeling

One of the issues in the early stage of mechatronic product design (i.e., system architecture) is to effectively derive the parameters of a product and its subsystems based on the abstract descriptions of products such as function requirements. To support system architecture, parametric representation (parameter network) using the definition of conceptual relations to identify the physical structure is crucial. The major challenge for the researchers studying conceptual design is to develop modeling schemes supporting the process to derive them. However, the modeling schemes should be independent of specific disciplines. In order to communicate with engineers from other disciplines, such as mechanical and control, a modeling scheme is therefore necessary for the system architects.

In this study, the FBS modeling scheme has been used as a fundamental solution to deal with the process of system architecture, because the FBS modeling scheme supports conceptual design independent of specific disciplines. The three key concepts that underpin FBS modeling are function, behavior, and state (Figure 1). A function is described as "a description of behavior abstracted by humans through recognition of the behavior in order to use it." It denotes a function as a combination of two ideas: symbol of human intention represented in the form of *to do something* and behavior that can exhibit the function. Behavior is

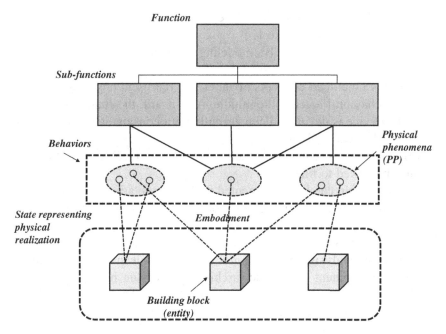

Figure 1. Hierarchical decomposition of system into functions, behaviors, and states using FBS modeling.

characterized as a series of states that change over time. A state is defined in this modeling by entities, their characteristics, and their structure, which also reflects the system's physical realization (Umeda *et al.*, 1996). Physical phenomena are used as symbolic concepts in the FBS modeling scheme to define (conceptual) relationships between product parameters.

FBS modeling is a domain-independent modeling scheme for representing a mechatronic system in a single model. In this model, function is an abstract term that can be applied to both hardware and software, as well as purely mechanical and electronically operated sensor actuator systems, and is independent of any domains.

FBS modeling is a design technique that enables the redesign and development of next-generation products by mapping function, behavior, and product structure. The embodiment of functions into physical form is also supported by FBS modeling.

Hierarchical decomposition of the system, which is based on the FBS modeling, is shown in Figure 1.

The advantages of the FBS modeling are as follows:

- It relates functional concepts of states that represent physical structure via a behavioral level in the middle. As a result, the functional model of the device is developed concurrently with consideration of the real-world physical environment.
- FBS modeling aids decision-making at the lower level, i.e., entities that must be linked to functions.
- This modeling can be implemented in the computational tool that supports system architecture tasks in complex mechatronic systems.

In the following section, a method is proposed and explained by comparative analysis of autonomous vacuum cleaning (VC) robots for the development of product architecture for the next generation's product.

4. A Method to Develop Product Architecture

In this study, a method (Figure 2) is proposed to support the development of product architecture for the next generation by (a) effectively utilizing the design knowledge of the current generation and by (b) computationally supporting the development process. A product modeling scheme and a computational modeling environment are required to demonstrate and validate the approach. The modeling system should, in particular, support domain-independent modeling descriptions and be consistent with a computational modeling environment. The study uses the FBS modeling and a CAD model for system architecture in a computational environment (SA-CAD).

This method illustrates the suitability of the scheme and computational support in the execution of the study. This method explains the application of the proposed method in the development of product architecture of autonomous vacuum cleaning robots. The proposed approach facilitates the creation of product platforms and product families based on FBS modeling of similar mechatronic systems. A brief schematic description of the proposed method is shown in Figure 2.

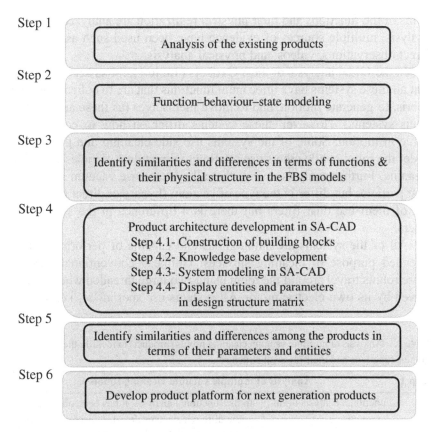

Step 1

Analysis of the existing products

Step 2

Function–behaviour–state modeling

Step 3

Identify similarities and differences in terms of functions &
their physical structure in the FBS models

Step 4

Product architecture development in SA-CAD
Step 4.1- Construction of building blocks
Step 4.2- Knowledge base development
Step 4.3- System modeling in SA-CAD
Step 4.4- Display entities and parameters
in design structure matrix

Step 5

Identify similarities and differences among the products in
terms of their parameters and entities

Step 6

Develop product platform for next generation products

Figure 2. Schematic description of steps in the proposed method. A method for the development of product architectures for next-generation products.

5. Results and Discussion

Results and the discussion of the proposed method are explained in the following.

5.1. *Step 1: Analysis of the existing products*

Initially, existing products of the same type are observed for their functionality. Next, the system designer identify the relationship between the function to structure. Then, the differences among the existing products in terms of the functions and their physical realization are analyzed. For this analysis, multiple sources of evidence have been used such as manuals, direct observations, videos, and physical analysis.

In the initial analysis of the VC robots (Table 1), it has been observed that all these systems have three main functions that are to collect dust and debris, to generate motion, and to move themselves (as these are autonomous systems). However, these systems differ on how to collect dust while in motion. Some of the systems use side cleaning and throw dust under the robot, while others use rotation to direct dust particles with side cleaning. Further analysis reveals that all systems use vacuum for lifting dust particles but differ in the type of vacuum (bypass or direct injection). All of them use dust filters but there is a difference in the size of the filters.

All of the systems are autonomous and capable of performing their intended purpose of cleaning surfaces. Unlike a conventional cleaner, these robots travel around the room using two large threaded wheels, each driven by its own electric motor. All systems use spot mode, i.e., to stay

Table 1.   Analysis of multiple vacuum cleaning robots (product) to identify the differences in their functions and their physical structure.

Step 1	Analysis of multiple vacuum cleaning robots	
a.	Functional analysis of all systems	In the functional analysis, it has been observed how the different systems detect, collect, and store the dust and particles while navigating autonomously. The three main functions identified are related to *collecting dust, generating motion, and moving itself.*
b.	Identification of functional differences	From functional analysis, differences in these systems are identified in sub-functions such as *to navigate itself and to collect dust and particles.*
c.	Physical analysis of all systems	Then, their physical structures are observed with criteria such as robot shape, size, number and type of sensors and actuators, way of collecting dust (i.e., using single or counter-rotating brushes), and way of lifting dust and particles.
d.	Differences in physical level of all robot systems	Differences in physical structure are identified in all systems on the basis of the above criteria.

at dirty spots, but differ in sensing used for dust particles. All systems follow walls and sense obstacles in their way and are also able to sense heights. They are able to perform self-charging when the battery is low and able to move back into their docking station and recharge themselves for the next time.

Further analysis reveals that there are some special functions in some systems, i.e., various onboard sensors and they are able to anti-tangle while stuck in rugs and cords. They are also able to navigate from one room to other room.

5.2. *Step 2: FBS modeling*

Following the analysis of existing products of the same kind, the proposed approach moves on to developing FBS models of multiple products. These models are developed with the following system architecture tasks in order to perform system decomposition and comparison:

- To model a product, customer requirements are converted into system-level specifications, i.e., abstract and parameter-level descriptions.
- The main function is decomposed into sub-functions.
- The system models are developed by transforming functions into behaviors and physical structures.

Using the system architecting tasks, FBS models of the VC robots are developed. For instance, in the FBS model, shown in Figure 3, functions are linked to physical phenomena (*PP*) at the behavior level and further linked to relevant entities at the state level. For instance, to fulfill the customer requirement such as cleaning floors independently, the main function is decomposed into sub-functions. For example function to navigate itself is decomposed into various sub-functions. One of the sub-functions, to avoid obstacles, is further decomposed into two sub-functions: to sense obstacles and to take turns. The robot senses obstacles by proximity and by touching. In the next step at the behavior level, physical phenomena are identified; for instance, to sense obstacles the designer identifies 'collision' as a physical phenomenon that is further related to physical entities. At the state level, this PP is further linked to the relevant entities. In this example, the physical phenomenon, i.e., collision, is related to object and bumpers as entities. Similarly, PP collision sensing is linked to

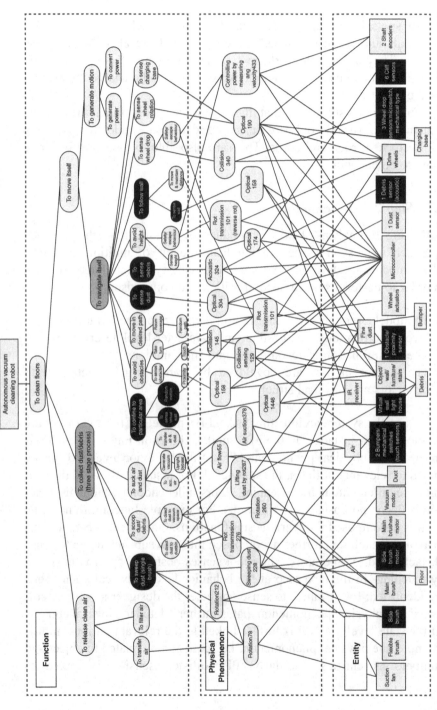

Figure 3. FBS model of robot system A. In the FBS model, customer requirements are transformed into system-level specifications and further into entities of the system.

mechanical switches (2-touch sensors), object and microcontroller, as entities. FBS models of the remaining systems are developed in the same manner.

5.3. *Step 3: Identify similarities and differences in FBS models*

FBS models of different products of the same type can be analyzed in this step to determine the differences in the respective systems. As illustrated in Figure 4, these models can be utilized to explicitly highlight similarities and differences in terms of functions and their physical decomposition. The differences between the models assist in determining how these systems might be redesigned.

Different shades are utilized to indicate the distinctions between the FBS models of the VC robots, as illustrated in Figure 3. These distinctions can be seen in functions (such as collecting dust and debris and navigating

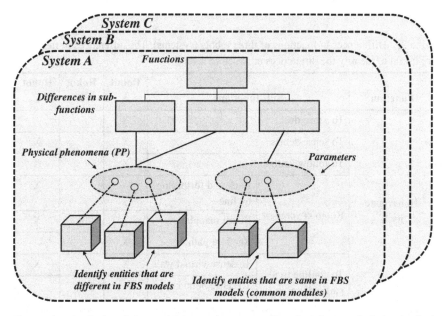

Figure 4. Analysis of the multiple products regarding functions and their physical realization.

itself), subsystems and components (such as the room positioning system and camera), and the environment in which they operate.

The functional decomposition reveals that the main functions of the three systems are same:

- To generate motion.
- To collect dust and debris.
- To release clean air.
- To navigate itself.

When the aforementioned functions are further broken into sub-functions, variations emerge in two of them: navigation and dust and debris collection. Because the physical manifestation of the functions to release clean air and generate motion is the same in all systems, they are not further divided into behaviors and states. The ability to navigate by oneself involves a variety of behaviors that each robot implements based on customer needs and capabilities, such as facilitating efficient navigation. Similarly, the function of collecting dust differs among the robots. Table 2 summarizes the functional differences among the three robots.

Table 2.   Differences in functions of three robots (systems). The differences in functions are useful to identify the differences at structure levels.

Function	Sub-functions		Robot A	Robot B	Robot C
To navigate itself	To sense dust		X		
	To sense debris		X		X
	To follow walls		X		X
	Room coverage	Back and forth: line by line		X	X
		Room mapping		X	X
		Random path	X		
	To confine to a particular area	To sense virtual wall	X		X
		To sense boundary markers		X	
To collect dust and debris	To sweep dust (one side)		X		
	To sweep dust (both sides)				X

FBS modeling support the designers to relate the functions as well as sub-functions to the relevant entities. For instance the sub-function i.e., to confine to particular area, robot A and robot C using optical sensors (entities) to sense virtual walls, while robot B using magnetic sensors (entities) to sense boundary markers. Similarly, objects, walls, dust, and debris can be linked to entities using the PP in between. The differences at the physical level can also be realized; however, this can be identified after the system architectures are formed.

5.4. *Step 4: Product architecture development in SA-CAD*

In this step, hierarchical physical decompositions at several levels (functional, behavioral, and structural) in SA-CAD are used to model each product. The architecture of the system is established as a network of parameters associated with distinct entities or building blocks realized by physical phenomena as a result of this modeling.

5.4.1. *Construction of building blocks for architecture modeling*

Mapping from conceptual relations to entities in the FBS models is employed in this sub-step to identify mapping from conceptual relations to customer requirements, as well as to obtain product design parameters and their dependencies. For instance, the vacuum motor, fan, and duct are the building blocks of the suction system in the VC example. Rotation, air suction, and airflow are their PPs. The relation between the duct and air is airflow, as shown in Figure 5.

In the following section, the design concepts and knowledge base development are described before architecture modeling.

The knowledge-intensive engineering framework (KIEF) supports a conceptual design process based on FBS modeling. Yoshioka (2004) defines it as a framework for integrating design modeling systems with embedded knowledge of domain theories. Several concepts such as physical phenomena, physical features, attributes of entities, and physical laws are described in the knowledge-intensive engineering framework and stored as data by the system architect.

This study uses the concepts of KIEF in FBS representation as follows:

Entity: An entity represents an atomic physical object and its purpose is to describe the abstract–concrete relationship among concepts. In this study, an entity is represented as a building block or module.

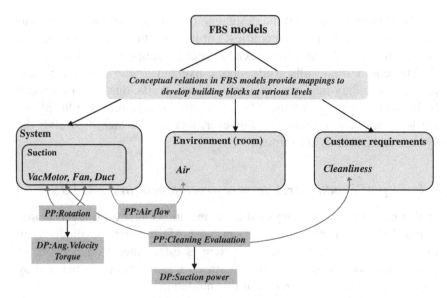

Figure 5.   An illustration of building blocks and their parameters for network modeling. The building blocks are related to relevant attributes.

Relation: A relation represents a relationship among entities to characterize a static structure.

Attribute: An attribute is a concept attached to an entity and takes a value to indicate the state of the entity.

Physical phenomenon: A physical phenomenon indicates physical laws or rules that govern behaviors.

Physical law: A physical law illustrates a simple relationship among attributes.

Behavior: Behavior is defined by sequential changes of states of a physical structure over time. For example, in the VC robot, the behavior of a vacuum motor depends on the torque generated.

State: States are the different modes of a physical system or entity.

5.4.2. *Knowledge base development at SA-CAD*

The parameters of the product and their relationships are represented by the network modeler in SA-CAD. The system architect uses CAD tool to create a knowledge base of physical phenomena, properties, entities, and their relationships (see Figure 6).

Referring to the VC robot example (Figure 6), cleaning evaluation is the PP, and cleanliness is the customer perception with regard to the ability of the system to clean floors; density and suction power are the other attributes that are entered for dust and vacuum motor, respectively.

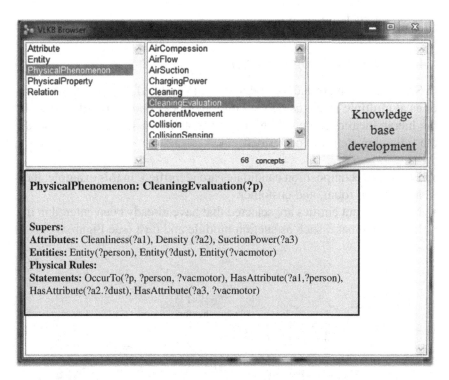

Figure 6.   An example of Knowledge base development in SA-CAD. A knowledge base including data of the concepts such as physical phenomenon (i.e., cleaning evaluation), attributes, entities, and their relationships is written in SA-CAD.

5.4.3 Metamodel development in SA-CAD

In SA-CAD, a metamodel represents a design object as a network of concepts, and these concepts are written and stored in the concept base of the tool. A metamodel is built as a network of these instantiated physical concepts. An important feature of this modeling representation is that, once the model is formed, any modification of the metamodel of the system is possible by either changing the existing concepts or adding new concepts later on.

For instance, in the VC example, an initial view of the metamodel is shown in Figure 7; this represents the concepts related to a function (i.e., cleanliness), along with physical phenomenon, attributes, entities, and their relations, which are all defined and written in SA-CAD. Ad data are extracted in the form of functions, physical phenomenon, and attributes to develop the metamodel.

5.4.4 System modeling in SA-CAD

In SA-CAD, each product is modeled by hierarchical physical decompositions at multiple levels using the building blocks and knowledge base. In the hierarchical decomposition of VC robots in SA-CAD (see Figure 8), the following steps are performed:

- First, the decomposition candidates are identified. In this example, they are system, room, and customer.
- Next, relevant entities are selected that have already been entered in the knowledge base, such as suction module and dust (see Figure 8).

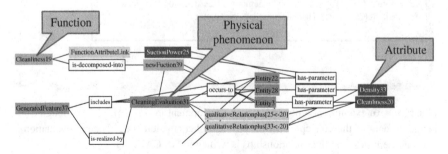

Figure 7.   An example of the metamodel in SA-CAD.

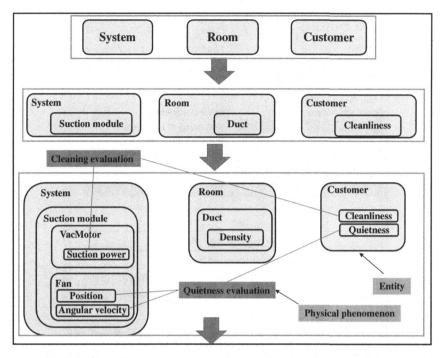

Figure 8.   Hierarchical system decomposition process of VC robot.

- Finally, the SA-CAD decomposition interface is used to relate the entities, their design parameters, and physical phenomena. For instance, the system architect defines *cleanliness* (as required by customer) as a function of the suction power of the vacuum motor in suction module. Here, entities (person and robot fan) and the design parameters (quietness, position, and angular velocity) are related parameters. Then, the architect relates the target parameter (i.e., quietness) and physical phenomena with the system elements (or entities) and executes the decomposition in SA-CAD as shown in Figure 9.

5.4.5 *System architectures as a parameter network*

The architecture of the system is formed as a network of parameters that are related to various entities or building blocks realized by the physical phenomena.

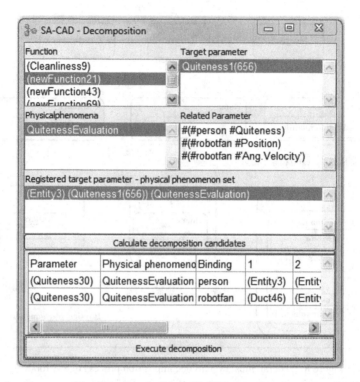

Figure 9.   Decomposition interface in SA-CAD. This interface is used to develop the parameter network in SA-CAD. Entity 3 represents the system element, quietness is a function, and quietness evaluation is the physical phenomenon defined in the decomposition interface.

The parameters of the product and their relationships are represented by the network modeler in SA-CAD. The system architect uses this tool to create a knowledge base of physical phenomena, properties, entities, and their relationships (see Figure 6). By using this method, multiple system architectures are developed (Figure 10) to facilitate product platforms and product family modeling. The commonality of the systems can be analyzed in the parameter network to find common building blocks.

The parameter network can also be used to identify the differences in the three VC robot architectures. The differences in their entities and

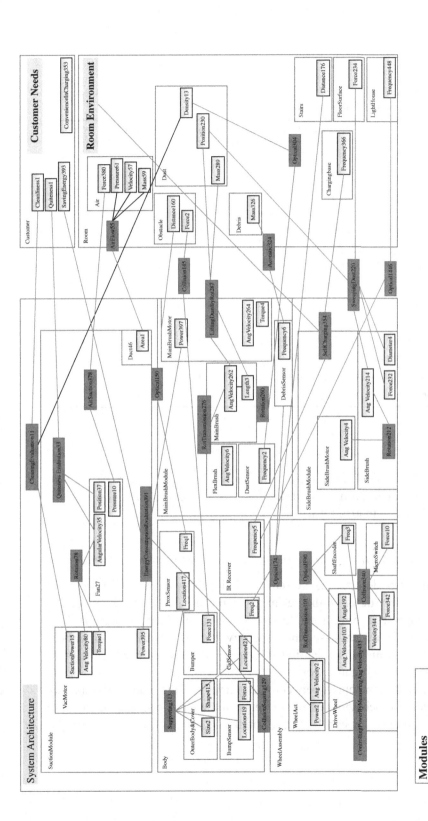

Figure 10.   Architectures of the VC robots in a parameter network relating physical phenomena, entities, and parameters or attributes.

Table 3.   Differences in the three architectures at behavior and structural levels. Numbers in each system represent the quantity of respective entities.

Entities	Physical phenomena	System A	System B	System C
Side brush	Rotation, sweeping dust	1	0	2
Side brush actuator	Rotation	1	0	2
Mechanical switch	Collision sensing	2	4	3
Light house	Optical	1	0	0
Proximity sensor	Optical	1	1	3
Debris sensor	Acoustic	1	0	1
Wheel drop sensor	Collision	3	2	0
Cliff sensors	Optical	6	2	3
Shaft encoder	Optical/Hall effect/ Coriolis effect	2	2	2
Magnetic sensors	Magnetic induction	0	2	0
Boundary markers	Magnetic induction	0	1	0
LASER sensor	Optical mapping	0	1	0
IR sensors	Optical	0	0	3

respective physical phenomena (are shown in Table 3) can be used to develop product families. Further, these parameters and entities can also be displayed in matrix form.

5.4.6 *Display entities and parameters in design structure matrix*

This sub-step depicts the system's relationships between entities and attributes. They describe the design concept established at the conceptual design stage. The tool assists designers in creating matrix representations (Figure 11) for matrix-based methodology (e.g., DSM) analysis of system architecture, which may be utilized for activities such as architecture creation and process organization.

The correspondences between the entities and the parameter relations that relate the behaviors of the systems to their respective structures are shown in Figure 11, which is based on the VC example.

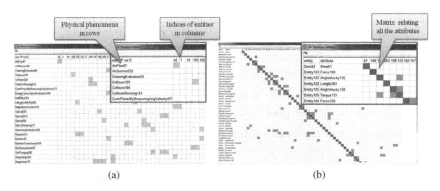

Figure 11. (a) Correspondences between entities and parameter relations in the VC robot example. These relations are derived from physical phenomena, and they illustrate the differences in these systems at behavioral levels. (b) Matrix representation of relations among attributes.

5.5. *Step 5: Identify similarities and differences among the products in terms of their entities*

The commonality of the three systems is manually detected in the parameter network in this step to find common building blocks or entities. Once the system's size is determined, the quantity and location of entities are determined based on the needs of various designs; this architectural distinction can be used to create product families.

For instance, in the parameter network (architectures) of the three VC robots (Figure 10), the entities and the associated physical phenomena in the suction system and the wheel assembly are almost the same. The only difference is the amount of sensors and where they are placed. For example, System A has six cliff sensors, System B has two, and System C has three (Table 4). Similarly, the main brush and main brush motor are shared by all three systems in the main brush module. Only System A and System C share the side brush module.

The architectural distinction can be used to create product families. As indicated in Table 4, this distinction is evident in functions such as collecting dust and debris and navigating itself, as well as their related entities.

Table 4.   Differences in entities of three VC robots (systems).

	Entities	System A	System B	System C
1	Main brush module	Two counter-rotating brushes	Single rotating brush	Single rotating brush
		Dust sensor		
		Debris sensor		Debris sensor
2	Side brush	Single side brush		Two side brushes
		Single motor		Two motors
3	Body	IR receiver		IR receiver
				Virtual wall sensor
		One proximity sensor	One proximity sensor	Three proximity sensors
				Camera
			Magnetic sensor	
4	Room mapping system		IR laser sensor	
			RPS motor	
			IR receiver	

5.6. *Step 6: Product platform development*

The proposed method could find multiple levels of classifications in terms of the level of similarity observed in current systems, including behavioral (physical phenomena) and structural levels (types of components (entities) used, as well as their numbers and locations). The similarities and differences discovered by such classifications have been utilized to construct product platforms. When effectively implemented, product platforms can provide the following benefits: organizations may produce distinctive products by sharing components across a platform, they can minimize development time, cost, and system complexity, and they can gain the flexibility to change to upgrade and redesign products.

In the VC case, two platforms are formed that have common entities based on the robots' round and D-shaped designs (Figure 12). These platforms are also extended to product families, allowing for the development of derivative products based on the three architectures (Table 5).

In the VC example, only two derivative products are shown for each platform (Figure 12). However, it is possible to have more derivative products based on the following:

- Number and type of sensors, for instance, the number of cliff sensors can be varied.
- The number of actuators varies depending on the needs; for instance, the D-platform has less actuators than the round platform.
- Type of vacuum (i.e., direct or bypass).
- Way of lifting dust and debris.

With the development of product platforms and families, companies can offer differentiated products to customers. With common and differentiated modules in the architectures, industries develop their production lines based on common parts and variable parts. This reduces the cost and complexity, and improves flexibility in the manufacturing process.

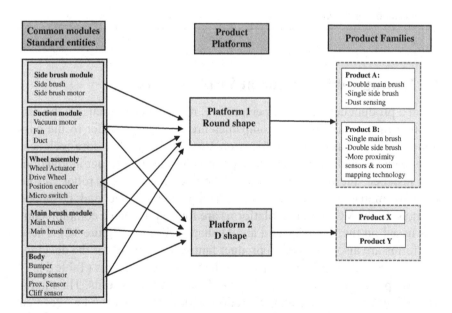

Figure 12. Product platforms based on common modules of the three systems' architectures.

Table 5. Product families (based on two platforms) can be developed using the differentiation in the product architectures (refer to Tables 3 and 4).

Differentiation in entities in the three architectures	Platform 1		Platform 2	
	Product A	Product B	Product X	Product Y
Two counter-rotating main brushes	■			
Single rotating main brush		■	■	■
Single side brush	■			
Two side brushes		■		
Dust sensor	■			
Debris sensor		■		
Virtual wall sensor		■		
Single proximity sensor	■		■	■
Three proximity sensors		■		
IR receiver	■			
Camera		■		
Single magnetic sensor			■	
Two magnetic sensors				■
Room positioning system			■	■

6. Managerial and Social Implications

From a product development perspective, managers can focus on the distinctive functionalities their companies are pursuing to offer to the customers. Thus, managers can focus on strengthening their core technical competencies and R&D focus in support of initiating product innovation. For instance, the modular design in the vacuum cleaning robot can be used by managers to develop two versions, i.e., round and D-shaped designs. In the VC case, two platforms are formed that have common entities based on the robots' round and D-shaped designs (Figure 12). These platforms are also extended to product families, allowing for the development of derivative products based on the three architectures (Table 5).

Adopting a modular design approach has several benefits. The modular design enables easier administration. More resources can be allocated at the beginning of product development. This approach enables parallel product development that leads to easier communication and administration at multiple levels. Modular architecture and platforms can also be used by firms to structure the manufacturing process. By identifying the architecture and required interfaces between the modules, the assembly process and the required manufacturing resources can be identified early in the design stage. Thus, the managers can foresee and design manufacturing setup at the conceptual design stage. For instance, in the VC robot case, once the basic architecture including actuators, sensors, and controllers was designed, managers could decide to undertake concurrent developments of the modules in different departments in the company to accelerate product development.

By applying a modular approach in conjunction with technology innovation in design, companies can identify a manufacturing setup and develop derivative products for the markets that accelerate the innovation process.

Mass customization approaches have social implications as well. For example, modular product design (MPD) affects human diversity by bringing a variety of products that fit requirements of different groups. This approach can facilitate involvement of customers in the design stage, i.e., identify the product performance from the market and incorporate the features that customers demand. MPD can affect customer ethics through customization of product features, uses, or cost. From the cost perspective, the modular approach has a positive impact on both the supply chain and product development cost savings when certain production activities are performed simultaneously.

7. Conclusion

Designers and engineers have the critical role of designing and optimizing modules and their interfaces in the early stages of multidisciplinary products. Computational support is vital for the effective execution of this task. In addition, the data accumulated during the system design can be used for R&D in product and process innovation.

This study proposed a method for supporting this task using FBS modeling and the SA-CAD tool. The FBS models were developed with the system architecture tasks and divided into four processes. First, customer requirements were converted into system-level specifications. Second, system-level specifications were transformed into subsystems and components. Third, the subsystems and components were realized with their desired functions and behaviors. Fourth, the system performance was analyzed. The interaction between the components and the design parameters was identified.

After the FBS models, the differences and similarities between the models were identified. The hierarchical physical decompositions at several levels (functional, behavioral, and structural) in SA-CAD were used to model each product. The architecture of the system was established as a network of parameters associated with distinct entities or building blocks realized by physical phenomena as a result of this modeling.

In SA-CAD, each product is modeled by hierarchical physical decompositions at multiple levels using the building blocks and knowledge base. In the CAD tool, the VC robot architectures were developed. The parameter network was used to identify the differences in the three VC robot architectures. The differences in their entities and respective physical phenomena were used to develop product families.

The similarities and differences identified by such classifications have been utilized to construct product platforms. In the VC case, two platforms were formed that have common entities based on the robots' round and D-shaped designs (Figure 12). These platforms were also extended to product families, allowing for the development of derivative products based on the three architectures. The proposed method would enable the development of several products in a product family and identification of components that are shared by products in a product family to provide designers with flexibility to deal with drastically changing market needs.

In short, this method supports the task of system decomposition, complexity management, interface development, and the identification of system properties at multiple levels in a multidisciplinary product.

References

Alvarez Cabrera, A. A., Foeken, M. J., Tekin, O. A., Woestenenk, K., Erden, M. S., De Schutter, B., *et al.* (2010). Towards automation of control software: A review of challenges in mechatronic design. *Mechatronics, 20,* 876–886.

Burr, J. (1990). A theoretical approach to mechatronic design. PhD thesis, Denmark: Technical University of Denmark.

Berman, B. (2002). Should your firm adopt a mass customization strategy? *Business Horizons, 45*(4), 51–60.

Colombo, E. F., Shougarian, N., Sinha, K., *et al.* (2020). Value analysis for customizable modular product platforms: Theory and case study. *Research in*

Engineering Design, 31, 123–140. https://doi.org/10.1007/s00163-019-00326-4.

Chmarra, M. K., Cabrera, A. A., van Beek, T., D'Amelio, V., Erden, M. S., & Tomiyama, T. (2008). Revisiting the divide and conquer strategy to deal with complexity in product design. In *IEEE/ASME International Conference on Mechatronics and Embedded Systems and Applications,* 2008 Oct 12 (pp. 393–398). IEEE.

Dieterle, W. (2005 Jan 1). Mechatronic systems: Automotive applications and modern design methodologies. *Annual Reviews in Control,* 29(2), 273–277.

Dahmus, J. B., Gonzalez-Zugasti, J. P., & Otto, K. N. (2001). Modular product architecture. *Design Studies,* 22, 409–424.

Fredriksson, P. & Gadde, L. E. (2005 Oct 1). Flexibility and rigidity in customization and build-to-order production. *Industrial Marketing Management,* 34(7), 695–705.

Hehenberger, P. (2009 Jan 1). Application of mechatronic CAD in the product development process. *Computer-Aided Design and Applications,* 6(2), 269–279.

Habib, T. & Komoto, H. (2014 Oct 1). Comparative analysis of design concepts of mechatronics systems with a CAD tool for system architecting. *Mechatronics,* 24(7), 788–804.

Habib, T., Kristiansen, J. N., Rana, M. B., & Ritala, P. (2020 Dec 1). Revisiting the role of modular innovation in technological radicalness and architectural change of products: The case of Tesla X and Roomba. *Technovation,* 98, 102163.

Hvam, L., Mortensen, N. H., & Riis, J. (2008). *Product Customization.* Springer Verlag, Edition, Vol. 1, pp. 17–40.

Hsiao, S. W. & Liu, E. (2005). A structural component-based approach for designing product family. *Computers in Industry,* 56, 13–28.

Komoto, H. & Tomiyama T. (2010). A system architecting tool for mechatronic systems design. *Ann CIRP,* 59(1), 171–174.

Komoto, H. & Tomiyama, T. (2012 Oct 1). A framework for computer-aided conceptual design and its application to system architecting of mechatronics products. *Computer-Aided Design,* 44(10), 931–946.

Liu, Z., Wong, Y. S., & Lee, K. S. (2010 Jun 15). Modularity analysis and commonality design: A framework for the top-down platform and product family design. *International Journal of Production Research,* 48(12), 3657–3680.

Mikkola, Juliana, H. (2007). Management of product architecture modularity for mass customization: Modelling and theoretical considerations. *IEEE Transactions on Engineering Management,* 54(1).

Martin, M. V. & Ishii, K. (2002). Design for variety: Developing standardized and modularized product family architectures. *Research in Engineering Design,* 13, 213–235.

Nielsen, K., Brunoe, T. D., Joergensen, K. A., & Taps, S. B. (2014). Mass customization measurements matrics. *Proceedings of the 7th World Conference on Mass Customization, Personalization and Co-Creation (MCPC 2014)*. Aalborg, Denmark.

Pahl, G. & Beitz, W. (2008). *Engineering Design: A Systematic Approach*. London: Springer.

Pine, B. J. (1993). *Mass Customization: The New Frontier in Business Competition*. Boston, MA: Harvard Business School Press.

Stone, R. & Wood, K. (2000). Development of a functional basis for design. *Journal of Mechanical Design*, 122(4), 359–370.

Steffen, N. J. (2013). Developing modular manufacturing system architectures: The foundation to volume benefits and manufacturing system. Changeability and responsiveness. PhD thesis, Aalborg University Denmark.

Tomiyama, T., D'Amelio, V., Urbanic, J., & El Maraghy, W. (2007). Complexity of multidisciplinary design. *Ann CIRP*, 56(1), 185–188.

Umeda, Y., Ishii, M., Yoshioka, M., & Tomiyama, T. (1996). Supporting conceptual design based on the function–behavior–state modeler. *Artificial Intelligence for Engineering Design, Analysis and Manufacturing*, 10(4), 275–288.

VDI 2206. (2004). Design handbook 2206. In *Design Methodology for Mechatronic Systems*. Düsseldorf: VDI Publishing Group.

Walcher, D. & Piller, F. (2012). The customization 500: A global benchmark study of online B to C mass customization (1st ed.). www.mc-500.com.2012.

Yoshioka, M., Umeda, Y., Takeda, H., Shimomura, Y., Nomaguchi, Y., & Tomiyama, T. (2004). Physical concept ontology for the knowledge intensive engineering framework. *Advanced Engineering Informatics*, 18, 95–113.

Chapter 4

Business Model Innovation Analytics for Small to Medium Enterprises

Matheus Franco[*,¶]**, Vinicius Minatogawa**[†,||]**, Ruy Quadros**[*,**]**,
Orlando Duran**[‡,††]**, Antonio Batocchio**[§,‡‡]**, Jose Garcia**[†,§§]
and Matias Valenzuela[†,¶¶]

[*]*Department of Science and Technology Policy, Geosciences Institute,
University of Campinas, Campinas, Brazil*
[†]*Escuela de Ingeniería en Construcción, Pontificia Universidad
Católica de Valparaíso, Valparaíso, Chile*
[‡]*Mechanical Engineering School, Pontificia Universidad
Católica de Valparaíso, Quilpué, Chile*
[§]*School of Mechanical Engineering, University of Campinas,
Campinas, Brazil*
[¶]*matheusfranco400@gmail.com*
[||]*vinicius.minatogawa@pucv.cl*
[**]*ruy@unicamp.br*
[††]*orlando.duran@pucv.cl*
[‡‡]*batocchi@fem.unicamp.br*
[§§]*jose.garcia@pucv.cl*
[¶¶]*matias.valenzuela@pucv.cl*

Hypercompetition requires shorter life cycles for products, services, processes, and business models. Innovation management aided by innovation analytics is helping companies respond more precisely to such challenges. Business model innovation, yet, is poorly explored in combination with innovation analytics. The subject is even more neglected when approached under the SME perspective. Such companies struggle in dealing with scarce resources. Still, they are also exposed to hypercompetitive environments. A theoretical background covering the evolution of innovation management and the importance of business model innovation-specific practices is provided. Thus, we will introduce a framework that enables SMEs to utilize innovation analytics. Presenting a comprehensible process, we integrate dynamic capabilities, activities, and data analytics. The aim is data-driven focused business model innovation. As for theoretical implications, this is one of the first studies relating business model innovation and innovation analytics. In practice, it will help managers to better understand innovation analytics for further application in SMEs.

1. Introduction

Hypercompetition (D'Aveni & Gunther, 2007) requires shorter product cycles, as it increases global competitiveness and the degree of uncertainty (King, 2013). In this scenario, achieving and sustaining competitive advantages is harder, because they tend to become obsolete rapidly due to the fast changing environment (Mahto *et al.*, 2018). Consequently, the same notion of fast change and the need to constantly innovate apply to products, services, processes, and Business Models (BM).

This hypercompetition phenomenon gained strength with Digital Transformation and the changes in the competitive business landscape (Verhoef *et al.*, 2018). The increase in uncertainty resulting from the digital transformation injected more turbulence in the business landscape (Schoemaker *et al.*, 2018). Even though the effect of digital transformation is not homogeneous throughout different industries, it is safe to say that its pervasiveness impacted almost every economic activity (Wiggins & Ruefli, 2005; D'Aveni *et al.*, 2010), pushing organizations toward the need to embrace digital transformation.

Digital transformation involves an organization's ability to adapt or promote new business models (Schallmo & Williams, 2018; Verhoef *et al.*, 2021) emphasizing the critical relevance of creating capabilities for

Business Model Innovation (BMI) (Franco *et al.*, 2021). BMI is a funda-mental process on different spectrums. The literature refers to BMI as an essential mechanism for conferring organizational sustainability (Minatogawa *et al.*, 2020) and increased performance (Franco *et al.*, 2021; Johnson *et al.*, 2008; Teece, 2010). BMI is also an essential element for technology-led innovations, such as the ones within industry 4.0 (Ibarra *et al.*, 2018). In addition, its role has been discussed in industries strongly affected by the current COVID-19 pandemic crisis (Harms *et al.*, 2021; Priyono & Moin, 2020; Breier *et al.*, 2021).

Achieving BMI, however, is still a considerable challenge. Many BMI attempts fail (Geissdoerfer *et al.*, 2018; Christensen *et al.*, 2016), causing severe economic consequences for companies (Chesbrough, 2007), delay-ing, for example, the adoption of sustainable solutions (Geissdoerfer *et al.*, 2017). Large companies can use their slack resources to sense and seize opportunities for new business models. Cases like BASF (Winterhalter *et al.*, 2017) demonstrate that such companies can allocate entire depart-ments to innovation, impacting its BMI capabilities, being somewhat less sensible to economic drawbacks from failures and increasing the applica-bility of trial-and-error approaches for creating BMI capabilities.

Regarding Small and Medium Enterprises (SMEs), the challenge is more significant. Few SMEs make an effort for BMI and most of those fail to get expected results (Latifi *et al.*, 2021). Usually, SMEs are likely to focus on addressing the current BM's emerging issues, neglecting the exploration of new opportunities (Kesting & Günzel-Jensen, 2015). However, exploiting only their current BM to achieve economic effective-ness, which alone is challenging, presents a risk in the current hypercom-petitive reality.

1.1. *The need for a data-driven approach for BMI*

Despite the growth in the BMI literature, the state-of-the-art approach for conducting the BMI process is through experimentation and trial and error (Silva *et al.*, 2019; Minatogawa *et al.*, 2019; Cosenz & Bivona, 2021). Many studies demonstrated the relevance of experimentation capability as a path to cope with uncertainty and increase the BMI process's effective-ness (Konietzko *et al.*, 2020; Baldassarre *et al.*, 2020; Weissbrod & Bocken, 2017; Bocken & Geradts, 2020; Ma & Hu, 2021; Bocken & Snihur, 2020). However, it is still an often-random process with many more failures than successes (Christensen *et al.*, 2016; Minatogawa *et al.*,

2019). This poses a significant problem for SMEs since every fail, besides inflicting costs, also might reduce the willingness for future BMI efforts, leading to risk aversion.

For SMEs, the BMI experimentation process needs to be more assertive. It needs to be data driven, making decision as accurate as possible and reducing results' randomness. As data science evolves, so does its contribution to improve innovation management processes, particularly in regard to sensing and seizing capabilities (Lin & Kunnathur, 2019). Netflix's "House of Cards" series is a case in point, as it used analytics to define, for example, the cast and script (Lin & Kunnathur, 2019; Carr, 2013; Mazzei & Noble, 2017).

Analytics is a technological solution for data-driven decision-making, and has recently been explicitly applied to innovation, creating the so-called "innovation analytics", with applications in the fuzzy front end of innovation (Kakatkar *et al.*, 2020). Innovation analytics is a recent approach with little literature basis. When considering BMI analytics specifically, there is no literature on the subject to the best of the authors' knowledge. As noted, however, this type of innovation can have a profound impact. Hence, studies that aid companies, especially SMEs, with data-driven approaches, will reduce subjectivity and randomness, thus calling for a data-driven approach for SMEs developing BMI.

In this chapter, the authors will contribute by providing a framework that integrates data science aspects for decision-making in the business model innovation process. Such contribution will comprise data science approaches for leveraging the dynamic capabilities of a company.

This chapter is subdivided into sections. In Section 2, we offer a theoretical background that will support the construction of the framework, addressing the problem situation raised in the introduction. Section 3 designs the framework, dividing it into different parts of the process involving data science and business model innovations. Finally, Section 4 presents our conclusions about this conceptual study.

2. Theoretical Background

2.1. *The evolution of innovation management*

Innovation management as a research field has evolved alongside the very theoretical knowledge in innovation studies. Thus, the emerging concern with the management of business model innovation and the digital transformation of business models mirrors the substantial growth in research and literature focusing on business models and business model innovation. In this section, we present a concise overview of such evolution and argue that much of the growing interest in business models stems from the ever-growing importance of (digital) services and the challenges posed to the conventional business models of manufacturing industries by digitalization.

Let us get inspiration from Rothwell's seminal account of generations of innovation process models (Rothwell, 1994) for a brief synthesis of the evolution mentioned above. As the understanding of the innovation process leaves behind the linear model (Kline & Rosenberg, 1986; Rosenberg, 1982) and the need for coupling technology and market (Freeman & Soete, 2008) is largely accepted, the focus of innovation management moved from the dominant emphasis on invention and R&D management to the view of innovation management as a multifunctional process, integrating technology strategy and development strategy (Wheelwright & Clark, 1993). As the innovation dilemma was made clear (Christensen, 2000), that is, the problems posed by strongly distinctive capabilities and resources required from incremental and radical innovation, the understanding that external partners and networks are critical for successful innovation increased (Rothwell, 1994). In line with this, the literature on the management of innovation networks (Powell & Grodal, 2006) and later on Open Innovation (Chesbrough, 2003) has seen an enormous growth.

So far, the literature on innovation management, much like the most influential schools in innovation studies, namely, the neo-Schumpeterian and the evolutionary approaches, has largely focused on the manufacturing industry and the corollary of its business model, technological innovation of product and process (OECD, EUROSTAT, 2018). Thus, the most disseminated ideas and authors dealing with models, process, and tools focused on innovation management with an emphasis on technological innovation. At this point, it is important to bring in the French and the Dutch schools of studies on innovation in services (Gallouj & Savona, 2008; Bettencourt *et al.*, 2013; Oliveira & Von Hippel, 2011), whose authors early on disputed the adequacy of the manufacturing industry-inspired view of innovation to deal with services innovation. Perhaps their most relevant contribution was emphasizing the client as the central element of the innovation processes in services, as a critical actor. Moreover, some of their propositions for modeling innovation management in services deal with issues that are now explored in the literature in business model innovation such as value/service delivery and the modeling of services monetization (den Hertog *et al.*, 2010). Yet, so far, the innovation in services literature has been quite limited in changing the received wisdom about innovation processes and innovation management.

However, technical change is full of paradoxes, and it was up to the latest wave of disruptive and general-purpose technologies, i.e., digital transformation, to make even clearer the perception that (digital) services have become central to growth in the manufacturing (and non-manufacturing) industry and that innovation in business models assumed a central position in the debate on the innovation process. On the one hand, the potential scope for deep change in the current business model is vast, and it may affect most if not all functions in the firm. Digital transformation brings about potentially substantial changes in products, internal business processes, channels and links with clients and providers, forms of monetization, and the value proposition. On the other hand, learning from digital transformation can facilitate the entry into new markets, with new value proposals and entirely new businesses. In connection with this, we are dealing with a focus change, from product/process innovation to (digital services) business innovation. Last, but not least, as argued in this chapter, digital technologies have the potential to transform the consolidated processes in innovation management such as technology and competitive intelligence and ideation.

To conclude, the emergence of business model innovation opens up new opportunities, indeed, but we are also dealing in a terrain of limited experience, in which many of the new management practices will require trial-and-error attempts before being well understood and consolidated. Therefore, it is important to address and discuss the concept of business model innovation.

2.2. *What is business model innovation?*

Considering the purpose of this chapter of discussing business model innovation analytics, it is first crucial to understand the business model innovation concept. As important as it is to get a definition of the term, we also need to understand what characterizes business model innovation. What characteristics can one observe in a new business model that differentiates and makes it innovative? Although the discussion on these questions may be to some extent subjective, it's clarification is essential to later aggregate the analytics concept.

The concept of business model innovation has evolved considerably in the last 20 years (Foss & Saebi, 2016). Mitchell and Coles (2004) understood BMI as BM changes that provide product and service offerings to end customers that were not previously available. Gambardella and McGahan (2010) proposed the idea that business model innovations occur when a company takes a new approach to commercializing its assets. Yunus *et al.* (2010) argue that it is not only about creating new sources of revenue but also about making new value propositions.

Other authors relate business model innovation with business logic issues. Bucherer *et al.* (2012) define the term as a process that deliberately alters the central elements of a company's logic and business logic. Along the same lines, Casadesus-Masanell & Zhu (2013) understand that BMI refers to the companies' search for new logic and new alternatives to create and capture value for all the stakeholders, focusing mainly on finding new ways to generate revenue and defining value propositions for customers, suppliers, and partners. Eppler & Hoffmann (2013) declare that BMI is a multiphase process by which organizations transform new ideas into improved business models, seeking to advance, compete, and differentiate successfully in the market.

Thus, BMI is directly related to the systemic search for innovative ways of designing the flow of value. In this regard, two approaches may be considered. On the one hand, it considers an existing business model where innovation seeks to improve and increase its efficiency. On the other hand, completely new business model innovation seeks to create new markets. Examples cited by Johnson *et al.* (2008) illustrate these distinct processes well. There is the example of Hilti, a company that sells tools to the construction industry and whose business model consisted direct sales of products to its customers. However, the organization realized that its consumers generated value by delivering the contracts, not by owning the tools. With this, a business model innovation process resulted in a service offering, with the proposal to rent the tools instead of selling them according to the customers' need. This change significantly increased the effectiveness of Hilti's Business Model.

According to Schneider & Spieth (2013), although highly relevant to companies, especially those operating in volatile environments, existing knowledge about how organizations prepare to explore opportunities through innovations in business models is limited. They also state that it is necessary to achieve a deeper understanding of the factors that lead to the opportunity to innovate in business models in response to this situation. Similarly, Foss & Saebi (2016) also emphasize the need of studies that demonstrate the capability, leadership, and learning mechanisms necessary for successful business model innovations to occur.

The creation of new business models, however, is a process surrounded by subjectivity. It is still conceived ad hoc in many instances (Christensen *et al.*, 2016). Adopting the same product and service innovation processes does not necessarily result in new business models (Franco *et al.*, 2021; Tidd & Bessant, 2018). In this context, companies need to develop not only ordinary capabilities but also dynamic capabilities as a means to enable this type of innovation to emerge.

2.3. *Dynamic capabilities for BMI*

Resources are the basis for any successful process, and, when combined to make solid routines and capabilities, they create value (Winter, 2003). Capability is defined as "high-level routine (or collection of routines) that together with its implementing input flows, confers upon an organization's management a set of decision options for producing significant outputs of a particular type" (Winter, 2003). Routines have their roots in acquired and standardized knowledge, creating processes that are realized repetitively.

This definition of capabilities and routines places them both as the way a company generates profits at a specific time point (Laaksonen & Peltoniemi, 2018). Digital technologies had an impact on competition dynamics, injecting turbulence and rendering many BMs uncertain. If innovation was a critical competitive element since the mid-19th century (Nelson, 2007), especially technological innovation, today the vast possibility of proposing a business around a value proposition, and the ease in doing so in a low-cost way, has gained even more momentum and become a must for almost every company. This increase in turbulence means that solid capabilities and routines that promote profit become obsolete faster and faster (Schoemaker *et al.*, 2018).

How to live in a competitive world where capabilities and routines that underpin a BM can become obsolete at any time? The answer is that we need dynamic capabilities for changing our mainstream capabilities in a systematic fashion. With dynamic capabilities, a company can adapt and be mutable in a turbulent world. Hence, the conceptual distinction between dynamic capabilities and ordinary capabilities is relevant (Laaksonen & Peltoniemi, 2018). Ordinary capabilities are responsible for the operational excellence of a current BM — how do we make money now? Dynamic capabilities are responsible for reshaping the ordinary ones — how do we keep making money in the future? Bringing this capability discussion into business model language, high-level ordinary capabilities are the pillars that sustain the value creation, delivery, and capture architectures. Dynamic capabilities, by definition, change the value creation, delivery, and capture architectures and are, hence, the primary BMI antecedent.

What are dynamic capabilities? As they are tailored for changing something, they look like any innovation or improvement process. They are separated into three major parts, sensing, seizing, and transforming (Teece, 2018). Sensing means to have antennas looking anywhere for opportunities, which can be flaw in the existing process, finding new suppliers, new customers, and new technological applications. Seizing capability means to translate opportunities into solutions that become products or services and new BMs. Transforming capability means to escalate the BM to grow and compete in the market (Teece, 2007). The analogy looks like a funnel: More opportunities are coming in than new BMs coming out. Sensing, seizing, and transforming are the capabilities that execute the BMI process and are, therefore, the focus of one's attention when proposing innovation analytics for BMI. But first, the authors dive into data analytics.

2.4. *Data analytics*

Data science and the concept of analytics are gaining momentum in literature. Data analytics refers to technologies and processes of acquiring deep knowledge and extracting information from data (Cao, 2017). This topic is highlighted mainly due to some contemporary contexts and specific conditions: first, the condition of the great availability of data, bringing up the concept of big data (Choi, 2018); second, technique development for exploring the big data context. As new approaches emerged, such as machine learning, new opportunities have also arisen (Agarwal & Dhar, 2014).

Data analytics allows latent information extraction from raw data (Hadi *et al.*, 2018). This has great value since it enables, for example, one to identify trends. The combination of data analysis and IoT systems helps one understand such reality didactically. The main objective of data analytics in IoT systems is to interpret the data better and thus formulate effective decision-making (Saleem & Chishti, 2019). As an example, sensors can provide a large amount of data about the behavior of structures such as bridges and buildings (Kim & Queiroz, 2017). From Data Analytics, monitoring this type of structure can change paradigms. Maintenance that was previously carried out preventively can now be done in a predictive or prescriptive way (Zonta *et al.*, 2020).

It is important to highlight that trend identification is only one example made possible by the advances in machine learning techniques. Other important decision-making processes can also be assisted by such resources. Illustrating another type of example, in manufacturing, it is possible to use machine learning techniques for the detection of bottlenecks in the dynamics of a production system (Subramaniyan *et al.*, 2020).

Thus, it is observed that structured data can foster analytics, that is, data that already have the content organized, usually by means of rows, columns, and related tables for analysis. This structured nature facilitates search and storage, usually through relational database systems (Bennett *et al.*, 2019). Web Analytics, such as the Google Analytics tool, presents features where structured data are presented relating to websites (Li *et al.*, 2021). This type of analytics captures data that provide information such as demographics, location, devices, browsing behavior, and consumption habits, among many other resources.

However, data can also be unstructured. That is, specific resources that are not organized for data analysis can be structured to extract information. These features can be characterized as text files, audio files, and videos (Zhang *et al.*, 2019). Thus, there is an even broader context of Analytics applications if we also consider the feasibility of using unstructured data (Subramaniyaswamy *et al.*, 2015).

For instance, healthcare analytics has explored this wide variety of data (e.g., medical images, biomedical signals, audio transcripts, and handwritten prescriptions) to become a data-driven industry. In a more specific way, these resources help the decision-maker in the health industries answer various types of questions, depending on the type of analytics used.

Thus, descriptive, diagnostic, predictive, prescriptive, and discovery analytics can be presented (Mosavi & Santos, 2020).

In addition, it is also essential to consider the velocity at which data are made available. Such a feature makes a difference, for example, for social media analytics. Analyzing social networks typically covers the idea of real-time data (Gu *et al.*, 2016) and reporting incidents to emergency response systems is labor-intensive. We propose to mine tweet texts to extract incident information on both highways and arterials as an efficient and cost-effective alternative to existing data sources. This paper presents a methodology to crawl, process and filter tweets that are accessible by the public for free. Tweets are acquired from Twitter using the REST API in real time. The process of adaptive data acquisition establishes a dictionary of important keywords and their combinations that can imply traffic incidents (TI). This is critical for entertainment decision-makers. In this industry, it is necessary to quickly understand your audience's reaction to a particular event and/or content.

Thus, three features can be highlighted, volume, variety, and velocity. These are the characteristics that are commonly used to define the concept of Big Data Analytics (Saggi & Jain, 2018). The examples cited are didactic to exemplify each of the characteristics. However, it is important to note that volume, velocity, and variety are inherent in all examples of analytics cited.

Nevertheless, something little explored in the literature is how analytics can be exploited for decision-making regarding business model innovation. As already discussed, this type of innovation is extremely relevant for different types of industries and companies. There may be the argument that this is simply a variation of innovation analytics. However, analogous to product and service innovation processes, which have already been proved unsuitable for business models, the same could be argued for innovation analytics.

Thus, it is relevant to discuss theoretical bases that allow one to establish a clear analytics process for business model innovation. How could the business model innovation process be combined with this entire data context, which provides us with information and adds new knowledge? What approaches could be considered efficient, and at what stage of the process would it be more efficient? This discussion will lay the foundation for the proposition of a framework for Business Model Innovation Analytics.

3. Toward a Business Model Innovation Analytics Framework

3.1. *Framework proposal*

The BMI process can be understood as having four steps: (1) recognizing opportunities; (2) ideation and experimenting to refine opportunities and design potential solutions; (3) design and experiment with a prototype BM; and (4) implement and refine the BM (Frankenberger *et al.*, 2013; Geissdoerfer *et al.*, 2017). The sound execution of the BMI process depends on a set of dynamic capabilities, which are idiosyncratic of each company, and are a consequence of their BM structure, their decision-making processes, the set of activities and practices deployed for BMI, and the resources the company has available (Franco *et al.*, 2021).

SMEs have some advantages when we consider that their BM is considered more flexible than larger companies. Hence, they tend to display less friction and inertia for BMI (Franco *et al.*, 2021; Christensen *et al.*, 2016). Translating into the capability's language, SMEs have fewer BM structural constraints, which is positive. However, SMEs do not enjoy abundant resources. Instead, they usually have limited human, financial, organizational, and relational resources (Minatogawa *et al.*, 2020; Bouncken & Fredrich, 2016). Considering that the BMI process is resource-consuming and highly uncertain, it is understandable that SMEs show aversion to proactively pursuing BMI, if not necessary, due to threats to survival.

How can analytics help cope with uncertainty, reduce costs, and increase the effectiveness of SMEs' BMI efforts without the need for highly specialized resources? As we know, analytics, machine learning, and algorithms are more effective in knowledge-intensive activities than creative ones (Kakatkar *et al.*, 2020). Creative processes still depend on human activity, even with the advances in non-supervised machine learning and self-programming diverging algorithms. This idea is crucial in order to design a framework for BMI analytics effectively, once knowledge-intensive parts are highly dependent on resources (Denicolai *et al.*, 2014; Berends *et al.*, 2016).

With that said, before proposing BM innovation analytics, we need to break down the BMI process to understand which parts are more knowledge-intensive and which ones are more creative-intensive. To this end, we combine the double-diamond innovation process with the abovementioned BMI process.

It is initiated by identifying an opportunity and demanding in-depth knowledge about a specific customer's needs, which are still unmet and may be a latent or non-identified need. This process leads to the ideation and creative process of selecting and defining which opportunities have potential solutions, and seeking to create the so-called problem–solution fit pairs (Ries, 2011; Blank, 2007; Johnson, 2010), the very base of a new BM. This stage is a creative process in which activities are primarily deployed to improve creativity, which is also the core idea of the supporting tools and practices such as Design Thinking and ideation through business model visual tools. After ideating, the outputs are potential problem–solution pairs, which call for experimentation to generate knowledge — another knowledge-intensive stage.

After validating the problem–solution pairs, the next step is to design a potential BM, which also comprises the associated products and/or services, and the definition of the intended target customers' market. In this stage, several potential BMs are designed, characterizing this as a creative-intensive process. Next, there is an experimentation process to validate the best BM design, testing the financial model to see if it provides returns and check what the value creation and delivery architectures should be. This is also a knowledge-intensive process.

Finally, one important highlight is that the process is not linear or continual, as the experimentation stages often iterate with the creative design stages. Finally, when introducing the BM into the market, the process moves forward to traditional strategic management, focusing on creating efficiency and BM evolution. Figure 1 depicts the double-diamond BMI process, separated between creative- and knowledge-intensive stages, indicating the innovation analytics role as potentializing the knowledge-intensive parts of the process. It is important to highlight that every pair of knowledge-intensive and creative-intensive stage is iterative — the process is not linear and has many return points.

In each knowledge-intensive stage, innovation analytics acts as a supporting activity to refine knowledge and lower the cost of access to knowledge. Hence, it has a data-driven decision-making aspect but does not replace human activity by any means. The effective deployment of innovation analytics for BMI depends on answering crucial questions:

Business Model Innovation Process

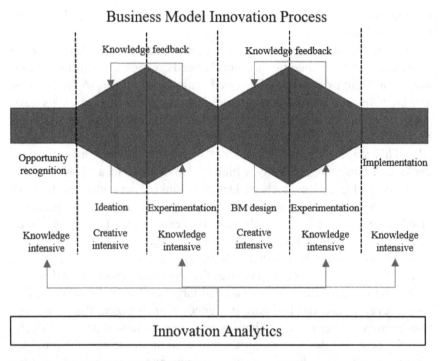

Figure 1.   The Business Model innovation process and its key stages. The innovation analytics proposed to optimize the knowledge-intensive parts of the process.

- Data collection:
  - Why collect data? (Improve performance, refine customer segmentation, discover new customers, explore new technologies, explore new markets, etc.)
  - What data? (Customers, internal processes, other markets, new technologies, etc.)
  - Where to gather data? (Customers, special purpose experiments, enterprise resources' planning systems, internal databases, internal spreadsheets, etc.)
  - Who will collect the data? (Which teams, formed by which members, with which skill sets)
  - How will we collect the data? (Leverage built-in analytics such as Google analytics, Facebook data, design surveys, data crawlers, natural language processing in social media, etc.)

- Data storage:
  o Where to store the data? (Cloud, internal servers, special databases, SQL, or NoSQL)
  o Who will store the data? (Internally develop resources, externally managed through partners)
  o Which data will be stored? (Filter the irrelevant from the relevant data — closely linked to the data collection strategy and strategic BMI goal)
  o How to store data? (Create database capabilities, delegate to external partners)
- Data analysis:
  o Who will explore and how to explore the data? (Exploratory Data analysis techniques)
  o Which data analysis technique to use? (K-Nearest neighbors, Regressions, K-means-clustering, supervised or non-supervised machine learning, deep learning, etc.)
  o Who will analyze the data? (Data scientists, internally trained employees, external partners)
  o How to apply data analysis into the BMI process? (Data visualization techniques and teams' dynamics)

3.2. *Framework usage*

The core idea of applying innovation analytics to SMEs' BMI efforts is to reduce resource consumption while improving performance and allowing for ambidextrous behavior. The main drawback is that BMI efforts are highly context-dependent and specific to each company's problems. Searching for a plug-and-play solution is hardly a productive effort. We do not aim to provide such a solution. Instead, our goal is to help SMEs create capabilities to successfully decide how to apply innovation analytics to their BMI efforts.

Bearing this in mind, it is hard to foresee a scenario where SMEs will be able to deploy business model innovation analytics without at least a data scientist or a few data scientists, depending on the company's size. It means that some specialized human resources will be necessary. Even though it may sound costly, with the advancement of information and communication technology, computer power on the one hand and the evolution of programming languages such as Python and R, with sophisticated libraries for data collection, cleaning, and analysis, on the other, this is increasingly feasible for SMEs with low resources.

Considering the challenges surrounding BMI, and the abovementioned issues, we propose three core recommendations: (1) It is important to move out of the inertia and begin experimenting with the BMI process. The best way of doing this is to begin with less uncertain efforts, focusing on the current BM issues that can provide good performance increase returns. This process is the basis for evolving to more advanced BMI efforts, since it works as a virtuous cycle by improving motivational levels and freeing resources. (2) One should integrate data scientists into the multidisciplinary BMI teams, thus not only enhancing creativity levels but also improving the innovation analytics and the BMI process and making it sophisticated. (3) One should solidify capabilities for BMI, creating new organizational structures and processes as to make this activity a capability and not an isolated event.

3.2.1. *Recommendation 1: Start small but start*

Considering this need for specialized resources and the need to amplify existing resources, our roadmap for business model innovation analytics begins by focusing on gaining performance and developing digital capabilities. Hence, a good start is to diagnose the current business model stage to find its key weaknesses, which are the high-leverage points. Deploy the first data scientists to assist in this diagnosis stage. Following the abovementioned questions, one can gather data from the company's customer segments, using social media, promoting surveys, profiling their behavior for improving value proposition, and increasing the enticement to pay for the company's services and/or products. Next, use this stage as a growth spiral, as performance gains in the current BM create financial resources through economic gains, human resources, learning by doing, and experience with digital capabilities, while also gaining experience with executing the BMI process.

It is a co-evolution between learning and creating capabilities to execute the business model innovation process with the creation of digital innovation capabilities. Naturally, data science and innovation analytics alone will not lead to practical performance gain results. Data science and business model innovation management walk hand in hand. Innovation management and business model innovation knowledge are essential for extracting success from BMI analytics. Therefore, patience and strategic focus may be the key elements for achieving success.

3.2.2. *Recommendation 2: Build multidisciplinary teams*

The BMI process alone relies upon multidisciplinary teams working together to leverage creativity and providing the technical and marketing bases of a BM. Usually, BMI efforts are conducted under project management logic, following a more agile paradigm. Hence, the teams should include people with complementary skills, relevant for designing and implementing the BM. As such, it should involve technical, marketing and sales people. Finally, dealing with novel BMs means that many customers and markets are unknown, and integrating external stakeholders into the process helps in the agile creation of knowledge to improve the BMI process.

The integration of innovation analytics means adding data scientists to the BMI teams. This integration requires constant knowledge exchange, since the technical and marketing knowledge relevant for the BM guides the creation of adequate questions for data scientist to answer. Algorithms alone will certainly not deliver high-performance results: they need good and relevant questions if they want to provide valuable answers.

Hence, it is particularly important to explore recommendation 1, beginning with more feasible BMI, which helps the teams gain knowledge and experience with the BMI process. With data scientists working together, this also improves the knowledge about how to use data, leading to more sophisticated data analysis and further improving the effectivity of BMI. Partnership with data-oriented companies and universities may also help cover the gaps and weaknesses in data analytics without increasing costs by hiring many specialists. Figure 2 shows the co-evolution between recommendations 1 and 2, focusing on the interplay between the BMI process knowledge and data science capabilities. Finally, it depicts the iterative nature between both recommendations.

3.2.3. *Recommendation 3: Rethink structure, solidify BMI analytics capability*

BMI is costly, uncertain, and may be painful, but the rewards are equally substantial. Beginning the process and acquiring knowledge of how to

Initiate and evolve

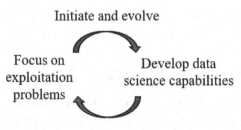

Current BM diagnosis	Identify data needs
• Prioritize problems with most potential financial returns • Refine customer segmentation • Test different value capture mechanisms • Refine internal value creation processes • Understand core value proposition	• Hire/train data scientists • Explore data-oriented partners • Explore partnerships with universities • Identify infrastructural needs – servers, computers, among others.

Figure 2. Recommendation to begin applying the BMI process on current BM performance improvement and co-evolve the BMI capability with the digital innovation capability.

execute the BMI process are very challenging. But, using innovation analytics on costly activities may help reduce these costs while also reducing uncertainty, which can have powerful effects on the acceptability of incorporating BMI efforts as a part of the business.

Similar to the mid to late 19th century, when innovation, particularly technological innovation, became part of business, today, we BMI is almost a must, given the new competitive dynamics brought about in the digital era. Thus, incorporating BMI as a new business function is essential. In the first half of the 20th century, companies created the R&D function, which was integrated into the overall BM. BMI changes the BM, affecting strategic elements, which may cause problems when proposing a similar solution. We propose changing the company structure to include BMI as a new function at the highest level possible, such as in the C-Level.

Doing so allows for continual deployment of activities and practices for BMI and BMI analytics efforts. This leads to a positive learning curve to assist companies in incorporating this capability into their BM. Hence, we suggest focusing on how this should be done, considering each company's BM logic, as a one-size-fits-all solution may not work.

4. Concluding Remarks

This chapter aims to provide a framework that integrates data science aspects into decision-making with regard to the business model innovation process. Considering that the literature on the innovation analytics subject is very recent, there is a scarcity of literature that provides appropriate theoretical bases for business model innovation analytics.

Within such a context, it was opportune to develop a framework that would be able to guide decision-making regarding business model innovation — especially considering SMEs and the characteristics inherent to these kinds of organizations. Large companies can use their slack resources to develop teams that integrate different data science and innovation management skills. In spite of the discussion on the potential rigid differences between SMEs' BMs and large companies' BMs, it may be somewhat advantageous for SMEs to embark into the BMI world.

To fulfill the objective of the chapter, a theoretical construction was proposed. Initially, the concept and importance of the theme of innovation were discussed. The myriad of possibilities and innovation types enables a range of specific solutions. Innovation Analytics is focused more on product, process, and service innovation. Business model innovation has a different complexity compared to these other types of innovation. Thus, just as innovation management processes need to be adapted to business models, it is coherent to argue that innovation analytics also needs to consider particular aspects for these archetypes.

An interesting opportunity stems from the existing gap around the systematization of the BMI process. Hence, the emergence and need to understand the subject related to dynamic capabilities. Such capabilities differ from routine activities, that is, ordinary capabilities. Dynamic capabilities are linked to activities such as sensing, seizing, and transforming.

The authors argue that dynamic capability can be combined with data analytics. The big data context and the volume, variety, and velocity characteristics bring new possibilities for data-driven decision-making, especially when combined with techniques developed for this context. Thus, interesting results are already presented in different areas of knowledge, branching out to descriptive, diagnostic, predictive, and prescriptive analytics.

This chapter, therefore, presented a framework toward business model innovation analytics. The steps of the business model innovation process and the activities of dynamic capabilities were merged. In the framework, important questions relevant to data collection, data storage, and data analysis were presented. Finally, the authors established recommendations on the use of the framework proposal considering the context of the SMEs.

It is not clear whether a framework with the characteristics and purpose as the one presented in this chapter is available in the literature. As discussed previously, BMI is a key factor enabling digital transformation. This paper, however, noted the possibility that this relation may present a two-way street. Digital transformation enhances a company's ability to create and analyze data to make decisions about BMI.

Although it is a theoretical proposition, the present study contributes to the literature by aggregating this combination of innovation, business models, dynamic capability, and data analytics. From a practical point of view, there is an interesting contribution for managers and decision-makers of SMEs, since the framework can operate as a support in the data-driven decision-making process for business model innovation.

Acknowledgment

The authors acknowledge funding provided by DI Interdisciplinaria Pontificia Universidad Católica de Valparaíso (PUCV), Valparaíso (PUCV), 039.414/2021.

References

Agarwal, R. & Dhar, V. (2014). Big Data, data science, and analytics: The opportunity and challenge for IS research. *Information Systems Research*, 25(3), 443–448.

Baldassarre, B., Konietzko, J., Brown, P., Calabretta, G., Bocken, N., Karpen, I. O., *et al.* (2020). Addressing the design-implementation gap of sustainable business models by prototyping: A tool for planning and executing small-scale pilots. *Journal of Cleaner Production*, 255, 120295. https://doi.org/10.1016/j.jclepro.2020.120295.

Bennett, R. M., Pickering, M., & Sargent, J. (2019 Apr). Transformations, transitions, or tall tales? A global review of the uptake and impact of NoSQL, blockchain, and big data analytics on the land administration sector. *Land Use Policy*, 83, 435–448.

Berends, H., Smits, A., Reymen, I., & Podoynitsyna, K. (2016). Learning while (re)configuring: Business model innovation processes in established firms. *Strategy Organization*, 14(3), 181–219. https://doi.org/10.1177/1476127016632758.

Bettencourt, L. A., Brown, S. W., & Sirianni, N. J. (2013). The secret to true service innovation. *Business Horizon*, 56(1), 13–22. http://dx.doi.org/10.1016/j.bushor.2012.09.001.

Blank, S. G. (2007). *The Four Steps to the Epiphany: Successful Strategies for Products that Win*. Palo Alto, CA: Cafepress.

Bocken, N. & Geradts, T. H. J. (2020). Barriers and drivers to sustainable business model innovation: Organization design and dynamic capabilities. *Long Range Planning*, 53(4) (October), 101950. https://doi.org/10.1016/j.lrp.2019.101950.

Bocken, N. & Snihur, Y. (2020). Lean Startup and the business model: Experimenting for novelty and impact. *Long Range Planning*, 53(4), 101953. https://www.sciencedirect.com/science/article/pii/S0024630119303887.

Bouncken, R. B. & Fredrich, V. (2016). Business model innovation in alliances: Successful configurations. *Journal of Business Research*, 69(9), 3584–3590. http://dx.doi.org/10.1016/j.jbusres.2016.01.004.

Breier, M., Kallmuenzer, A., Clauss, T., Gast, J., Kraus, S., & Tiberius, V. (2021). The role of business model innovation in the hospitality industry during the COVID-19 crisis. *International Journal of Hospitality Management*, 92, 102723. https://doi.org/10.1016/j.ijhm.2020.102723.

Bucherer, E., Eisert, U., & Gassmann, O. (2012). Towards systematic business model innovation: Lessons from product innovation management. *Creativity and Innovation Management*, 21(2), 183–198.

Cao, L. (2017). Data science. *ACM Computing Surveys*, 50(3), 1–42.

Carr, D. (2013). For "House of Cards," using Big Data to guarantee its popularity. *The New York Times*.

Casadesus-Masanell, R. and Zhu, F. (2013). Business model innovation and competitive imitation: The case of sponsor-based business models. *Strategic Management Journal*, 34(4), 464–482. http://dx.doi.org/10.1002/smj.2022.

Chesbrough, H. (2003). *Open Innovation: The New Imperative for Creating and Profiting from Technology*. 1st ed. Boston, MA: Harvard Business School Press.

Chesbrough, H. (2007). Business model innovation: It's not just about technology anymore. *Strategic Leadership*, 35(6), 12–17.

Choi, T. M., Wallace, S. W., & Wang, Y. (2018). Big Data analytics in operations management. *Production and Operations Management*, 27(10), 1868–1883.

Christensen, C. M. (2000). *The Innovator's Dilemma*. 2nd ed. New York: Harper Business.

Christensen, C. M., Bartman, T., & van Bever, D. (2016). The hard truth about business model innovation. *Sloan Management Review*, 58(1), 31–40.

Cosenz, F. & Bivona, E. (2021). Fostering growth patterns of SMEs through business model innovation. A tailored dynamic business modelling approach. *Journal of Business Research*, 130, 658–669. https://www.sciencedirect.com/science/article/pii/S0148296320301594.

D'Aveni, R. A. & Gunther, R. (2007). Hypercompetition. Managing the dynamics of strategic maneuvering. In *Das Summa Summarum des Management* (pp. 83–93). Gabler.

D'Aveni, R. A., Dagnino, G. B., & Smith, K. G. (2010 Dec). The age of temporary advantage. *Strategic Management Journal*, 31(13), 1371–1385.

den Hertog, P., van der Aa, W., and de Jong, M. W. (2010). Capabilities for managing service innovation: Towards a conceptual framework. *Journal of Service Management*, 21(4), 490–514.

Denicolai, S., Ramirez, M., & Tidd, J. (2014). Creating and capturing value from external knowledge: The moderating role of knowledge intensity. *R&D Management*, 44(3), 248–264.

Eppler, M. J. & Hoffmann, F. (2013). Strategies for business model innovation: Challenges and visual solutions for strategic business model innovation. In Pfeffermann, N., Minshall, T., & Mortara, L. (eds.), *Strategy and Communication for Innovation*. Berlin, Heidelberg: Springer Berlin Heidelberg (pp. 3–14). https://doi.org/10.1007/978-3-642-41479-4_1.

Foss, N. J. & Saebi, T. (2016). Fifteen years of research on business model innovation: How far have we come, and where should we go? *Journal of Management*, 43(1), 200–227. https://doi.org/10.1177/0149206316675927.

Franco, M., Minatogawa, V., Duran, O., Batocchio, A., & Quadros, R. (2021). Opening the dynamic capability black box: An approach to business model innovation management in the digital era. *IEEE Access*, 9, 69189–69209.

Frankenberger, K., Weiblen, T., Csik, M., & Gassmann, O. (2013). The 4I-framework of business model innovation: a structured view on process phases and challenges. *International Journal of Product Development*, 18(3/4), 249. http://www.inderscience.com/link.php?id=55012.

Freeman, C. & Soete, L. (2008). A economia da inovação industrial. In *Clássicos da Inovação*. Campinas: Unicamp (pp. 335–494).

Gallouj, F. & Savona, M. (2008). Innovation in services: A review of the debate and a research agenda. *Journal of Evolutionary Economics*, 19(2), 149. https://doi.org/10.1007/s00191-008-0126-4.

Gambardella, A. & McGahan, A. M. (2010). Business-model innovation: General purpose technologies and their implications for industry structure. *Long Range Planning*, 43(2–3), 262–271. http://dx.doi.org/10.1016/j. lrp.2009.07.009.

Geissdoerfer, M., Savaget, P., & Evans, S. (2017). The Cambridge business model innovation process. *Procedia Manufacturing*, 8(October 2016), 262–269. http://dx.doi.org/10.1016/j.promfg.2017.02.033.

Geissdoerfer, M., Savaget, P., Bocken, N. M. P., & Hultink, E. J. (2017). The circular economy — A new sustainability paradigm? *Journal of Cleaner Production*, 143, 757–768. https://www.sciencedirect.com/science/article/ pii/S0959652616321023.

Geissdoerfer, M., Vladimirova, D., & Evans, S. (2018). Sustainable business model innovation: A review. *Journal of Cleaner Production*, 198, 401–416. https://doi.org/10.1016/j.jclepro.2018.06.240.

Gu, Y., Qian, Z., & Chen, F, (2016 Jun). From Twitter to detector: Real-time traffic incident detection using social media data. *Transportation Research Part C: Emerging Technologies*, 67, 321–342.

Hadi, M. S., Lawey, A. Q., El-Gorashi, T. E. H., & Elmirghani, J. M. H. (2018). *Big Data Analytics for Wireless and Wired Network Design: A Survey. Vol. 132, Computer Networks*. Elsevier B.V. (pp. 180–199).

Harms, R., Alfert, C., Cheng, C. F., & Kraus, S. (2021). Effectuation and causation configurations for business model innovation: Addressing COVID-19 in the gastronomy industry. *International Journal of Hospitality Management*, 95(October 2020), 102896. https://doi.org/10.1016/j. ijhm.2021.102896.

Ibarra, D., Ganzarain, J., & Igartua, J. I. (2018). Business model innovation through Industry 4.0: A review. *Procedia Manufacturing*, 22, 4–10. https:// www.sciencedirect.com/science/article/pii/S2351978918302968.

Johnson, M. W. (2010). *Seizing the White Space: Business Model Innovation for Growth and Renewal*. Massachusetts: Harvard Business Press.

Johnson, M. W., Christensen, C. M., & Kagermann, H. (2008). Reinventing your business model. *Harvard Business Review*, 86(12), 57–68.

Kakatkar, C., Bilgram, V., & Füller, J. (2020). Innovation analytics: Leveraging artificial intelligence in the innovation process. *Business Horizon*, 63(2), 171–181.

Kesting, P. & Günzel-Jensen, F. (2015). SMEs and new ventures need business model sophistication. *Business Horizon*, 58(3), 285–293. http://www.science direct.com/science/article/pii/S0007681315000038.

Kim, Y. J. & Queiroz, L. B. (2017 Jun). Big Data for condition evaluation of constructed bridges. *Structural Engineering*, 141, 217–227.

King, B. L. (2013 Jul). Succeeding in a hypercompetitive world: VC advice for smaller companies. *Journal of Business Strategy*, 34(4), 22–30.

Kline, S. J. & Rosenberg, N. (1986). An overview of innovation. In Landau, R., & Rosenberg, N. (eds.), *The Positive Sum Strategy*. Washington, DC: National Academy of Press (pp. 275–305).

Konietzko, J., Baldassarre, B., Brown, P., Bocken, N., & Hultink, E. J. (2020). Circular business model experimentation: Demystifying assumptions. *Journal of Cleaner Production*, 277, 122596. https://www.sciencedirect. com/science/article/pii/S0959652620326433.

Laaksonen, O. & Peltoniemi, M. (2018). The essence of dynamic capabilities and their measurement. *International Journal of Management Reviews*, 20(2), 184–205.

Latifi, M.-A., Nikou, S., & Bouwman, H. (2021). Business model innovation and firm performance: Exploring causal mechanisms in SMEs. *Technovation*, 107, 102274. https://www.sciencedirect.com/science/article/pii/ S0166497221000559.

Li, X., Law, R., Xie, G., & Wang, S. (2021). *Review of Tourism Forecasting Research with Internet Data. Vol. 83, Tourism Management*. Elsevier Ltd. (p. 104245).

Lin, C. & Kunnathur, A. (2019 Dec). Strategic orientations, developmental culture, and big data capability. *Journal of Business Research*, 105, 49–60.

Ma, Y. & Hu, Y. (2021). Business model innovation and experimentation in transforming economies: ByteDance and TikTok. *Management Organization Review*, 17(2), 382–388. https://www.cambridge.org/core/article/business-model-innovation-and-experimentation-in-transforming-economies-bytedance-and-tiktok/057C8387EC953C14ED10AA8996F53947.

Mahto, R. V., Ahluwalia, S., & Walsh, S. T. (2018 Aug). The diminishing effect of VC reputation: Is it hypercompetition? *Technological Forecasting and Social Change*, 133, 229–237.

Mazzei, M. J. & Noble, D. (2017 May). Big data dreams: A framework for corporate strategy. *Business Horizon*, 60(3), 405–414.

Minatogawa, V. L. F., Franco, M. M. V., Rampasso, I. S., Anholon, R., Quadros, R., Durán, O., *et al.* (2019 Dec 30). Operationalizing business model innovation through Big Data analytics for sustainable organizations. *Sustainability*, 12(1), 277 (cited 2020 Feb 17). https://www.mdpi.com/2071-1050/ 12/1/277.

Minatogawa, V., Franco, M., Durán, O., Quadros, R., Holgado, M., & Batocchio, A. (2020). Carving out new business models in a small company through contextual ambidexterity: The case of a sustainable company. *Sustainability*, 12(6), 2337.

Mitchell, D. W. & Coles, B. C. (2004). Establishing a continuing business model innovation process. *Journal of Business Strategy*, 25(3), 39–49.

Mosavi, N. S. & Santos, M. F. (2020). How prescriptive analytics influences decision making in precision medicine. In *Procedia Computer Science.* Elsevier B.V. (pp. 528–533).

Nelson, R. R. (2007). Understanding economic growth as the central task of economic analysis. In Malerba, F. & Brusoni, S. (eds.), *Perspectives on Innovation.* Cambridge: Cambridge University Press (pp. 37–39).

OECD, EUROSTAT. (2018). *Oslo Manual 2018: Guidelines for Collecting, Reporting and Using Data on Innovation.* 4th ed. Paris/Eurostat, Luxembourg: The Measurement of Scientific, Technological and Innovation Activities, OECD Publishing.

Oliveira, P. & Von Hippel, E. (2011). Users as service innovators: The case of banking services. *Research Policy*, 40(6), 806–818.

Powell, W. W. & Grodal, S. (2006). Networks of innovators. In Fagerberg, J., Mowery, D. C., & Nelson, R. R. (eds.), *The Oxford Handbook of Innovation.* Oxford: Oxford University Press.

Priyono, A. & Moin, A. (2020). Identifying-digital-transformation-paths-in-the-business-model-of-smes-during-the-covid19-pandemic2020Journal-of-Open-Innovation-Technology-Market-and-ComplexityOpen-Access.pdf. *Journal of Open Innovation: Technology, Market, and Complexity*, 6(4), 104.

Ries, E. (2011). *The Lean Startup: How Today's Entrepreneurs Use Continuous Innovation to Create Radically Successful Businesses.* New York: Crown Books.

Rosenberg, N. (1982). *Inside the Black Box: Technology and Economics.* Cambridge: Cambridge University Press (p. 316).

Rothwell, R. (1994 Jan 1). Towards the fifth-generation innovation process. *International Marketing Research*, 11(1), 7–31. https://doi.org/10.1108/02651339410057491.

Saggi, M. K. & Jain, S. (2018 Sep). A survey towards an integration of big data analytics to big insights for value-creation. *Information Processing and Management*, 54(5), 758–790.

Saleem, T. J. & Chishti, M. A. (2019). Deep learning for internet of things data analytics. In *Procedia Computer Science.* Elsevier B.V. (pp. 381–390).

Schallmo, D. R. A. & Williams, C. A. (2018). *Digital Transformation Now! Guiding the Successful Digitalization of Your Business Model.* Cham, Switzerland: Springer International Publishing AG (p. 70).

Schneider, S. & Spieth, P. (2013). Business model innovation: Towards an integrated future research agenda. *International Journal of Innovation Management*, 17(01), 1340001. http://www.worldscientific.com/doi/abs/10.1142/S136391961340001X.

Schoemaker, P. J. H., Heaton, S., & Teece, D. (2018). Innovation, dynamic capabilities, and leadership. *California Management Review*, 61(1), 15–42.

Silva, D. S., Ghezzi, A., Aguiar, R. B. de, Cortimiglia, M. N., & ten Caten, C. S. (2019). Lean startup, agile methodologies and customer development for business model innovation: A systematic review and research agenda. *International Journal of Entrepreneurial Behavior & Research,* 26(4), 595–628.

Subramaniyan, M., Skoogh, A., Muhammad, A. S., Bokrantz, J., Johansson, B., & Roser, C. (2020). A generic hierarchical clustering approach for detecting bottlenecks in manufacturing. *Journal of Manufacturing Systems.* 55, 143–158.   https://www.sciencedirect.com/science/article/pii/ S0278612520300315.

Subramaniyaswamy, V., Vijayakumar, V., Logesh, R., & Indragandhi, V. (2015). Unstructured data analysis on big data using map reduce. In *Procedia Computer Science.* Elsevier B.V. (pp. 456–465).

Teece, D. J. (2007). Explicating dynamic capabilities: The nature and microfoundations of (sustainable) enterprise performance. *Strategic Management Journal,* 28(3), 1319–1350.

Teece, D. J. (2010). Business models, business strategy and innovation. *Long Range Planning,* 43(2–3), 172–194.

Teece, D. J. (2018). Business models and dynamic capabilities. *Long Range Planning,* 51(1), 40–49. http://dx.doi.org/10.1016/j.lrp.2017.06.007.

Tidd, J. & Bessant, J. (2018). Innovation management challenges: From fads to fundamentals. *International Journal of Innovation Management,* 22(5). DOI: 10.1142/s1363919618400078.

Verhoef, P. C., Broekhuizen, T., Bart, Y., Bhattacharya, A., Qi Dong, J., Fabian, N., *et al.* (2021). Digital transformation: A multidisciplinary reflection and research agenda. *Journal of Business Research,* 122(November 2019), 889–901.

Weissbrod, I. & Bocken, N. (2017). Developing sustainable business experimentation capability — A case study. *Journal of Cleaner Production,* 142(4), 2663–2676. http://dx.doi.org/10.1016/j.jclepro.2016.11.009.

Wheelwright, S. C. & Clark, K. B. (1993). *Managing New Product and Process Development.* New York: Free Press.

Wiggins, R. R. & Ruefli, T. W. (2005 Oct). Schumpeter's ghost: Is hypercompetition making the best of times shorter? *Strategic Management Journal,* 26(10), 887–911.

Winter, S. G. (2003). Understanding dynamic capabilities. *Strategy Management Journal,* 24(10 SPEC ISS.), 991–995.

Winterhalter, S., Weiblen, T., Wecht, C. H., & Gassmann, O. (2017 Apr 18). Business model innovation processes in large corporations: Insights from BASF. *Journal of Business Strategy,* 38(2), 62–75. https://doi.org/10.1108/ JBS-10-2016-0116.

Yunus, M., Moingeon, B., & Lehmann-Ortega, L. (2010). Building social business models: Lessons from the grameen experience. *Long Range Planning*, 43(2–3), 308–325. http://dx.doi.org/10.1016/j.lrp.2009.12.005.

Zhang, Y., Huang, Y., Porter, A. L., Zhang, G., & Lu, J. (2019 Sep). Discovering and forecasting interactions in big data research: A learning-enhanced bibliometric study. *Technological Forecasting and Social Change*, 146, 795–807.

Zonta, T., da Costa, C. A., da Rosa Righi, R., de Lima, M. J., da Trindade, E. S., & Li, G. P. (2020 Dec). Predictive maintenance in the Industry 4.0: A systematic literature review. *Computers & Industrial Engineering*, 150, 106889.

Chapter 5

Analysis of Factors Influencing Product and Process Innovation for Smart Manufacturing

Anilkumar Malaga[*] and Sekar Vinodh[†]

*Department of Production Engineering, National Institute of Technology,
Tiruchirappalli, Tamil Nadu, India
[*]aneelnitt@gmail.com
[†]vinodh_sekar82@yahoo.com*

Manufacturing is becoming sophisticated, computerized and complex. Intelligent production presents obstacles and opportunities. Implementation of Smart Manufacturing (SM) is a challenge for industry practitioners. The introduction of smart factories involves significant innovation in processes. However, organizations face enormous hurdles in introducing smart factories in view of the structural change that needs to be competitive. Process innovation is the introduction of new or significantly enhanced methods of development or distribution. The prospects for the technical and economic growth of smart products are enhanced by technological advancements. Today, in order to secure a significant growth force, the automotive sector strives to enhance productivity through product and process advancements in combination with the latest technologies. Product and process innovation has to be done in order to implement SM in organizations. Certain factors will influence innovation for SM. This work focuses on the identification of appropriate factors influencing product and process innovation in smart manufacturing. Factors are prioritized using an appropriate multi-criteria decision-making approach. The identification and analysis of influencing factors of product and process innovation for SM are the novel contributions of this study. The study enabled industry practitioners to focus on factors of product and process innovation for smart manufacturing.

1. Introduction

Implementation of smart factory is an invention phase. Process innovation is the introduction of modern or modified processes of manufacture or distribution. The automatic processes requiring few to no human interaction and even the systems that enhance communication between shop floor and market in real time create dynamical process inventions that interact with consumers and with other machines (Sjödin *et al.*, 2018). SM achieves manufacturing goals by interconnecting people, technologies, invention, and development (Malaga & Vinodh, 2021a). The main objective of the SM is to meet consumer requirements through research and production, recycle inventory, order processing, and agile implementation. The difficulty of the SM era is the sophistication of emerging technology that industrial companies need to embrace to introduce them successfully. This technical revolution is having a considerable effect on profitability, employment and development in the market, given the current scenario in the manufacturing industry (Wankhede & Vinodh, 2021).

SM is promoted by limited product and service life cycles and, therefore, by the desire to speed up market time. Complexity and expenditure may increase through innovation. It will also put organizations' capacity and capacity for innovation within the enterprise. SM is all about connectivity, including consumer integration, providers, and consumers. Internet of Things (IoT) and similar solutions and networks will use the collaboration layer to accelerate collaborative innovation. Clients and customers can be integrated through crowdsourcing techniques (Travaglioni *et al.*, 2020). In order for several types of real-time data to increase operational results and achieve their respective objectives, all smart manufacturing systems (SMS) operate through diverse channels and organizational models that cross or communicate with different segments of the industry.

The effects of digital technologies in production contribute to new diverse organizational challenges (Malaga & Vinodh, 2021b).

Production innovation is a typical area of research, and many studies evaluated the interconnection among company's wealth and its capacity to support a continued innovation phase (Marzi *et al.*, 2017). New interconnected and intelligent products revolutionize consumers' lives and can be called transformative technologies (Mani & Chouk, 2017). In the meantime, a promising demand for smart, integrated devices that can gather, interact, process, and deliver information has been enabled by the recent emergence of technologies (Zheng *et al.*, 2019). An evolution in the competitiveness of organizations that can expand or enter new markets is the product of innovation (Malaga & Vinodh, 2021a).

SM is focusing towards the satisfaction of consumer and manufacturer. For this, continuous innovation in manufacturing is essential in order to reach customers' changing requirements. Innovation can be done in products and processes in the manufacturing industry. Product and process innovation can be influenced by several factors. Hence, to attain successive inventions in product and process, the influencing factors have to be investigated. This forms the motivation of this study to identify and examine influencing factors. The study focused on the identification of the factors most influencing product and process innovation in SM of an automotive component manufacturing industry through literature review and validation by the expert panel. The identified factors have been prioritized using a Multi-Criteria Decision-Making (MCDM) approach.

The chapter is structured as follows: Section 2 covers the review of the literature and Section 3 focuses on the methodology. The conducted case study has been presented in Section 4, followed by results and discussion. Section 5 details the conclusion of this study.

2. Literature Review

Sjodin *et al.* (2018) explored the challenges in introducing intelligent facilities in the automotive industry and described essential measures required to incorporate an intelligent factory model. The authors developed a preliminary model of maturity to deploy smart factories' concept based on three key principles: digital cultivation, introduction of agile processes, and modular technology development. The authors concluded that the models enable realistic guidelines for applying intelligent manufacturing processes and translating manufacturing processes digitally, and contributing to the next wave of process creativity.

De Guimaraes *et al.* (2019) presented a framework for measurement and a systemic model to identify connections between the history of innovation and positive outcomes. The authors used a survey-based approach for the evaluation of the measurement of interrelations. The investigators concluded that the presented work facilitates decision-making in selecting deliberate drivers and investments in invention in order to attain competitive gain and financial benefits.

Larsen & Lassen (2020) examined the process of invention in design parameters which affect the novelty that results in intelligent manufacturing and consequently the design of invention processes for intelligent production. The investigators carried out action research that led to findings that explained a phenomenon in steady state and gained insight into how the phenomenon responds to changes. Furthermore, the authors stated that instruments and approaches that promote innovative practices that have a beneficial effect on innovation are missing.

Kahle *et al.* (2020) explored the characteristics needed to create intelligent goods through an environment of creativity. The authors established and revealed relationships between the conceptual structures which showed the features of an ecosystem invention to deliver intelligent products. The authors have concluded that ecosystem development allows SMEs to draw on complementary tools. It also allows developing country companies to address some of their shortcomings, but this would require more research into comparative cross-country analyses of certain ecosystems among developed and emerging countries.

Abdel-Basst *et al.* (2020) explored and presented an integrated model to assess innovation value propositions (IVPs) in smart product–service systems (SPSS) using the neutrosophic approaches of DEMATEL and CRITIC. The authors concluded that SPSS would significantly increase overall efficiency for all healthcare industry solutions: sensors, smart pills, and smart operations. The suggested system had been examined for the demonstrative study of smart continuous glucose surveillance sensors.

Travaglioni *et al.* (2020) explored and evaluated smart manufacturing in the open innovation context. The operational strategy is that the determinants define the phenomena to research interactions among them through the use of AHP and ISM methods. The authors concluded that other influences lead SM model despite being a poor catalyst in the open innovation view. Increasing their comprehension of these aspects and their interrelationships gives any stakeholder and decision-maker useful input into the SM, thereby fostering this openness model. *Research gaps*: Challenges in the introduction of SM in the automotive production industry have been explored (Sjödin *et al.*, 2018). Investigation of design parameters for process innovation has been done (Larsen & Lassen, 2020). The authors explored the drivers required to innovate smart products (Kahle *et al.*, 2020). Also, the authors examined SM in open innovation context (Travaglioni *et al.*, 2020). But the researchers have not focused on the significant factors that can influence product and process innovation in SM. Hence, the knowledge gaps are identified as the identification of significant factors influencing product and process innovation in SM and prioritizing the top factors in order to assist industry practitioners in product and process innovation for SM in the automotive component manufacturing industry.

3. Methodology

In this work, Fuzzy VIKOR multi-criteria decision-making approach is used as solution methodology. VIKOR tool provides compromise solutions through the collection of variables (Opricovic & Tzeng, 2004). The outcome which is closer to the optimal solution is referred to as a compromise solution. VIKOR is an effective tool of MCDM that prioritizes and sorts competing variables in order to optimize multiple complex attributes. VIKOR approach can completely reproduce decision-maker subjective desires, minimize individual regrets and maximize group utility.

The methodology followed in this study has been presented in Figure 1.

Figure 1 presents the steps of Fuzzy VIKOR approach used in this study. The detailed procedure of VIKOR approach is as follows (Vinodh *et al.*, 2014):

Stage 1: Describe the importance of the problem and the goal of decision-making.

Stage 2: Collection of appropriate criteria, factors, and members of the expert panel.

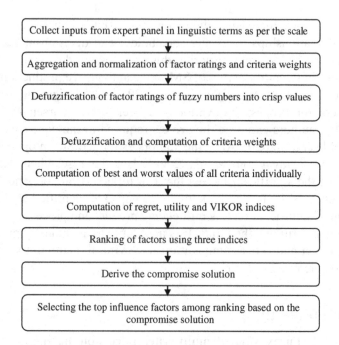

Figure 1.   Methodological steps of Fuzzy VIKOR approach.

Stage 3: Recognition of relevant linguistic parameters to weights of criteria and rating of attributes.

Trapezoidal fuzzy number sets have been used in this study. The linguistic scale for both criteria weights and ratings has been presented.

Stage 4: Aggregation and normalization of the criteria weight and attribute ratings using Equations (1)–(3) (Vinodh *et al.*, 2014).
Aggregation:

$$P_{ij} = \{P_{ij1}, P_{ij2}, P_{ij3}, P_{ij4}\} \tag{1}$$

where $P_{ij1} = \min_d (P_{ijd1})$, $P_{ij2} = 1/D (\sum_{(d=1)}^{D} P_{ijd2})$, $P_{ij3} = 1/D (\sum_{(d=1)}^{D} P_{ijd3})$, and $P_{ij4} = \max_d (P_{ijd4})$.
Normalization:

$$\mathfrak{U}_{ij} = \{P_{ij1}/P_{ij4}^{+}, P_{ij2}/P_{ij4}^{+}, P_{ij3}/P_{ij4}^{+}, P_{ij4}/P_{ij4}^{+}\}, Cj \in \mathfrak{B} \tag{2}$$

$$\mathfrak{U}_{ij} = \{P_{ij1}/P_{ij1}^{-}, P_{ij2}/P_{ij1}^{-}, P_{ij3}/P_{ij1}^{-}, P_{ij4}/P_{ij1}^{-}\}, Cj \in \mathbb{E} \tag{3}$$

where $P_{ij4}^{+} = \max_i \{P_{ij4}\}$, $C_j \in \mathfrak{P}$ ($\mathfrak{P}$: beneficial criteria set)

$$P_{ij1}^{-} = \min_i \{P_{ij1}\}, C_j \in \mathbb{E} \ (\mathbb{E}: \text{expense criteria set})$$

Similarly, the criteria weights also have to be aggregated and normalized.

Stage 5: Defuzzification of fuzzy numbers of ratings and criteria weights into crisp values using Equations (4) and (5).

$$\mathfrak{F}_{ij} = 1/4\{(P_{ij1}/P_{ij4}^{+}) + (P_{ij2}/P_{ij4}^{+}) + (P_{ij3}/P_{ij4}^{+}) + (P_{ij4}/P_{ij4}^{+})\}, \mathfrak{K}_j \in \mathfrak{P} \quad (4)$$

$$\mathfrak{F}_{ij} = 1/4\{(P_{ij1}/P_{ij1}^{-}) + (P_{ij2}/P_{ij1}^{-}) + (P_{ij3}/P_{ij1}^{-}) + (P_{ij4}/P_{ij1}^{-})\}, \mathfrak{K}_j \in \mathbb{E} \quad (5)$$

Stage 6: Assessment of all criteria best and worst values through Equations (6) and (7).

$$\mathfrak{F}_j^{*} = \max_i (\mathfrak{F}_{ij}) \quad (6)$$

$$\mathfrak{F}_j^{-} = \min_i (\mathfrak{F}_{ij}) \quad (7)$$

Stage 7: Assessment of indices: regret (R_i), utility ($\acute{S}_i$), and VIKOR (Q_i) are computed using Equations (8)–(10).

$$\acute{S}_i = \sum_{(j=1)}^{n} \{[W_j(\mathfrak{F}_i^{*} - \mathfrak{F}_{ij})]/[\mathfrak{F}_i^{*} - \mathfrak{F}_i^{-}]\} \quad (8)$$

$$R_i = \max_i \{[W_j(\mathfrak{F}_i^{*} - \mathfrak{F}_{ij})]/[\mathfrak{F}_i^{*} - \mathfrak{F}_i^{-}]\} \quad (9)$$

$$Q_i = [(v(\acute{S}_i - \acute{S}^{*}))/(\acute{S} - \acute{S}^{*}] + [((1 - v)(R_i - R^{*}))/(R^{-} - R^{*})] \quad (10)$$

where $\acute{S}^{-}$ = higher ($\acute{S}_i$), R^{-} = higher (R_i), $\acute{S}^{*}$ = lower ($\acute{S}_i$), R^{*} = lower (R_i), v is the highest utility, and $(1 - v)$ is the separate regret. From Wankhede and Vinodh (2021), the value of v is considered as 0.5.

Stage 8: Prioritization of factors depending on Q_i values. The factor having the smallest Q_i value is prioritized as first.

3.1. *Suggesting compromise solution*

To suggest the compromise solution, the following two conditions need to be satisfied (Opricovic & Tzeng, 2004):

Condition 1: Adequate profit $Q(R^2) - Q(R^1) \geq DQ$

where R^2 is the second place attained in the prioritization

$$DQ = 1/(\text{Number of factors} - 1)$$

Condition 2: Decision-making adequate stability. The factor ranked first should also be ranked first by utility and/or regret measures.

If the two conditions got satisfied, then the factor with the least index of VIKOR is the best factor, else more than one solution will be proposed as the best solution.

The notations used in this study have been summarized and presented in Appendix A.1.

4. Case Study

This study was conducted in an automobile components manufacturing industry in the Tamil Nadu state of India. The industry focused on product and process innovation in the context of SM. The factors affecting product and process innovation for smart production, which in turn allows industry to generate value, had to be identified. Industry practitioners and members of R&D with SM expertise have provided their opinions. The aim of this analysis is to consider and further analyze the factors by prioritizing through a suitable MCDM approach for efficient product and process innovation in the production of automotive components. Eighteen factors pertaining to SM are identified and analyzed against five criteria: Economy (C1), market attractiveness (C2), customer satisfaction (C3), product performance (C4), and operational excellence (C5). In this study, the authors identified 18 significant factors which influence product and process innovation for SM through the examination of existing research. The factors identified are human resources (F1) (Lu *et al.*, 2016; Gumbi & Twinomurinzi, 2020; Ghobakhloo, 2020), visualization (F2) (Sjödin *et al.*, 2018; Ali Qalati *et al.*, 2021), traceability of production (F3) (Horváth & Szabó, 2019), preventive maintenance (F4) (Horváth & Szabó, 2019), productivity and efficiency (F5) (Nimbalkar *et al.*, 2020), relative advantage (F6) (Gumbi & Twinomurinzi, 2020; Ghobakhloo, 2020; Ali Qalati *et al.*, 2021), cost effectiveness (F7) (Ali Qalati *et al.*, 2021; Lee *et al.*, 2017), agility (F8) (Larsen & Lassen, 2020; Bogle, 2017), collaboration (F9) (Sjödin *et al.*, 2018; Bogle, 2017), sustainability (F10) (Sjödin *et al.*, 2018; Zheng *et al.*, 2019), robustness (F11) (Bogle, 2017), security (F12) (Bogle, 2017), top management support (F13)

(Lin *et al.*, 2020), machine to machine communication (F14) (Lin *et al.*, 2020), compatibility (F15) (Gumbi & Twinomurinzi, 2020; Ali Qalati *et al.*, 2021), product re-configuration (F16) (Zheng *et al.*, 2019), competitive pressure (F17) (Ali Qalati *et al.*, 2021), and data integration (F18) (Lu *et al.*, 2016). The identified factors influence product and process innovation in relation to the corresponding technologies. During the implementation of SM technologies, these identified factors, which can influence product and process innovation, will come into picture. Identified factors influence product and process innovation from initiation of invention idea level until it reaches the customer. The factors are identified from the viewpoint of the customer, manufacturer and management. The identified factors can assist industry practitioners to attain successive inventions of products and processes. The panel includes experts from academia and the automotive components manufacturing industry with more than 15 years of experience in modern manufacturing. The identified factors have been validated by the expert panel. An MCDM technique named fuzzy VIKOR was employed for prioritization. Fuzzy VIKOR aims to prioritize the factors and offer a compromise solution that helps SM practitioners make appropriate decisions.

The criteria weights and factor ratings have been collected from members of expert panel in linguistic terms as follows (Girubha & Vinodh, 2012): factor ratings, such as very low (VL), low (L), fairly low (FL), medium (M), fairly high (FH), high (H), and very high (VH), and criteria weights, such as very poor (VP), poor (P), medium poor (MP), fair (F), medium good (MG), good (G) and very good (VG), are represented on a seven scale in trapezoidal fuzzy numbers such as (0.0,0.0,0.1,0.2), (0.1,0.2,0.2,0.3), (0.2,0.3,0.4,0.5), (0.4,0.5,0.5,0.6), (0.5,0.6,0.7,0.8), (0.7,0.8,0.8,0.9) and (0.8,0.9,1.0,1.0), respectively, for both ratings and weights. The linguistic terms of inputs have been received from experts and depicted in Table 1 for evaluating the indices.

Table 1 represents factor ratings corresponding to criteria and criteria weights' inputs from experts E1 and E2 on a seven scale in linguistic parameters (Girubha & Vinodh, 2012). The terms in Table 1 represent the linguistic terms as per the scale presented. Then, the linguistic terms have been translated into trapezoidal fuzzy numbers as per the scale.

The linguistic parameters have been transformed into fuzzy numbers and then aggregated and normalized the ratings and weights using equations (1)–(3). For this, the criteria pertaining to beneficial and non-beneficial categories are segregated. In this work, economy (C1) is

Table 1.   Expert input for the factor ratings and criteria weights.

Criteria	C1		C2		C3		C4		C5	
Factor	E1	E2	E1	E2	E1	E2	E1	E2	E1	E2
F1	H	M	FH	H	FH	H	VH	H	H	VH
F2	M	FH	L	M	FH	L	H	FH	FH	M
F3	FH	M	H	FH	FH	H	M	FH	M	FL
F4	H	FH	FH	M	FH	M	H	H	H	FH
F5	H	VH	H	FH	H	VH	H	VH	H	H
F6	H	M	H	FH	H	H	H	H	H	FH
F7	FH	M	M	FL	FH	M	M	FH	FH	M
F8	H	FH	H	H	H	VH	H	VH	H	VH
F9	FH	H	H	FH	H	H	H	H	H	FH
F10	M	FH	H	FH	VH	H	H	FH	FH	H
F11	FH	H	FH	M	H	VH	FH	M	H	VH
F12	H	FH	FH	H	H	H	H	H	FH	H
F13	H	H	H	VH	H	VH	H	VH	H	VH
F14	VH	H	FH	M	VH	H	VH	H	VH	H
F15	H	VH	M	FH	H	FH	H	FH	H	FH
F16	FH	H	FH	H	H	H	H	H	H	FH
F17	H	FH	H	VH	H	FH	H	H	H	H
F18	H	H	H	H	H	VH	VH	H	H	H
Weight	MG	G	G	MG	G	G	VG	G	G	VG

identified as non-beneficial criteria and the remaining four criteria are recognized as beneficial criteria. Then, the normalized fuzzy numbers have been defuzzified into crisp values using equations (4) and (5), and are depicted in Table 2.

Table 2 represents defuzzified factor ratings and criteria weights in crisp values. After aggregation and normalizing the factor ratings and criteria weights, fuzzy values have been defuzzified into crisp values and are depicted in Table 2.

Sample calculation:

For factor F1 corresponding to criteria C1, expert inputs are H and M.

Aggregation using equation (1) is = (0.4, 0.65, 0.65, 0.9).

Table 2. Defuzzification of factor ratings and criteria weights in terms of crisp values.

	C1	C2	C3	C4	C5		C1	C2	C3	C4	C5
F1	1.63	0.71	0.71	0.86	0.86	**F10**	1.47	0.71	0.86	0.71	0.71
F2	1.47	0.39	0.49	0.79	0.65	**F11**	1.43	0.59	0.86	0.59	0.86
F3	1.47	0.79	0.79	0.65	0.46	**F12**	1.43	0.79	0.89	0.89	0.79
F4	1.43	0.65	0.65	0.89	0.79	**F13**	1.14	0.86	0.86	0.86	0.86
F5	1.23	0.71	0.86	0.86	0.8	**F14**	1.23	0.59	0.86	0.86	0.86
F6	1.63	0.79	0.89	0.89	0.79	**F15**	1.23	0.65	0.79	0.79	0.79
F7	1.47	0.52	0.74	0.74	0.74	**F16**	1.43	0.79	0.89	0.89	0.79
F8	1.43	0.8	0.86	0.86	0.86	**F17**	1.43	0.86	0.71	0.8	0.8
F9	1.43	0.79	0.89	0.89	0.79	**F18**	1.14	0.8	0.86	0.86	0.8
						Weight	1.43	0.71	0.8	0.86	0.86

Table 3. Criteria for best and worst values.

	C1	C2	C3	C4	C5
Best $\mathfrak{F}$*	1.63	0.86	0.89	0.89	0.86
Worst $\mathfrak{F}^-$	1.14	0.39	0.49	0.59	0.46

Normalization using equations (2) and (3) is as follows:

$$\mathfrak{U}_{1,1} = (0.4/0.4,\ 0.65/0.4,\ 0.65/0.4,\ 0.9/0.4) = (1, 1.63, 1.63, 2.25)$$

$$\mathfrak{U}_{1,2} = (0.5/1, 0.7/1, 0.75/1, 0.9/1) = (0.5,\ 0.7,\ 0.75,\ 0.9)$$

Defuzzification of the normalization values which are converted into crisp values using equations (4) and (5).

For factor F1 and C1, $\mathfrak{F}_{1,1} = \frac{1}{4}(1 + 1.63 + 1.63 + 2.25) = 1.63$.

For factor F1 and C2, $\mathfrak{F}_{1,2} = \frac{1}{4}(0.5 + 0.7 + 0.75 + 0.9) = 0.71$.

Then, the overall performance has been analyzed through the computation of the best and worst values of criteria ratings from crisp values using equations (6) and (7). The computed best and worst ratings of all criteria have been depicted in Table 3.

Table 3 represents the computed best and worst values of all individual criteria from the defuzzified crisp values using equations (6) and (7).

4.1. Results and discussion

The purpose of this analysis is to prioritize product and process innovation factors for SM using fuzzy VIKOR for deriving solutions. The research

Table 4.   Indices of regret (R), utility ($\acute{S}$), and VIKOR index (Q).

Factor	$\acute{S}$	R	Q	Factor	$\acute{S}$	R	Q
F1	0.68	0.36	0.17	F10	1.6	0.52	0.42
F2	2.72	0.8	0.75	F11	1.91	0.86	0.61
F3	2.33	0.86	0.7	F12	0.84	0.58	0.29
F4	1.53	0.58	0.43	F13	1.58	1.43	0.77
F5	1.68	1.17	0.69	F14	1.73	1.17	0.7
F6	0.26	0.15	0	F15	2.13	1.17	0.78
F7	1.97	0.51	0.49	F16	0.84	0.58	0.29
F8	0.82	0.58	0.28	F17	1.33	0.58	0.39
F9	0.84	0.58	0.29	F18	1.8	1.43	0.81

focuses on prioritizing 18 SM-related factors with five criteria. This can be achieved using Equations (8)–(10) in calculating utility values ($\acute{S}$), regret measure (R) and VIKOR index (Q) and results depicted in Table 4. The final ranking depends on the least value of Q_i.

Table 4 presents utility, regret and VIKOR indices of identified factors influencing product and process innovation for SM. Three indices depicted in Table 4 have been computed using the best and worst values of criteria for all factors and defuzzified crisp values using equations (8)–(10).

The utility, regret and VIKOR indices of factors have been sorted in ascending order in order to obtain the final ranking depicted in Table 5. Sample calculations have been provided in Appendix A.2.

Table 5 represents the sequence of factors as per the ranking of three indices utility, regret and VIKOR.

4.2. *Compromise solution*

Compromise solutions have been provided based on the fulfillment of two conditions, i.e., acceptable benefit and decision-making adequate stability. Hence, two conditions need to be verified.

In this study, relative advantage (F6) attained the first rank and human resources (F2) attained the second rank as per the VIKOR index. Then acceptable benefit

Table 5.   Ranking sequence of factors based on $\acute{S}$, R, Q.

Rank	$\acute{S}_i$	R_i	Q_i	Rank	$\acute{S}_i$	R_i	Q_i
I	F6	F6	F6	X	F10	F17	F7
II	F1	F1	F1	XI	F5	F2	F11
III	F8	F7	F8	XII	F14	F3	F5
IV	F9	F10	F9	XIII	F18	F11	F3
V	F12	F4	F12	XIV	F11	F5	F14
VI	F16	F8	F16	XV	F7	F14	F2
VII	F17	F9	F17	XVI	F15	F15	F13
VIII	F4	F12	F10	XVII	F3	F13	F15
IX	F13	F16	F4	XVIII	F2	F18	F18

$$Q(\text{second rank}) - Q(\text{first rank}) \geq DQ$$

where $DQ = 1/(\text{number of factors} - 1)$

$$= 0.17 - 0 \geq (1/17) = 0.17 \geq 0.05$$

Hence, condition 1 got satisfied.

Condition 2 is adequate stability in expert's decision, i.e., the factor which attained the first rank as per the VIKOR index should also have attained the first rank as per the utility and/or regret.

Here, relative advantage (F6) attained the first rank by all three indices.

Hence, condition 2 is satisfied.

Hence, the factor with the least VIKOR index is the solution. Hence, relative advantage is the factor that can highly influence product and process innovation for SM.

5. Conclusion

In order to satisfy customer changing demands, product and process innovation has to be focused. SM initiates innovation in products and processes which are used in smart manufacturing systems and is essential for being competitive in the global market. The factors influencing SM product and process innovation have been identified, which gives more benefit to the industry practitioners to attain success through the deployment of SM in the industry. To fulfill this, the study focused on identifying factors influencing SM product and process innovation. In this context, the study contributed in identifying 18 influencing factors pertaining to product and process innovation for SM in an automotive component manufacturing industry. The study evaluated the influencing factors through the fuzzy VIKOR method. The fuzzy VIKOR approach facilitated the prioritization of factors and also provided compromise solutions. In this study, the top factor "relative advantage" (F6) influences more on the invention of SM products and processes. The ranking among the influencing factors facilitates industry practitioners to attain successive development in product and process invention in smart manufacturing systems. This study assists industry practitioners in the ease of identification of factors influencing product and process innovation for SM. Industry practitioners can focus more on the top-ranked factors in order to attain successive innovation in products and processes for SM. Through this analysis, industry practitioners can minimize capital investment and time to produce new products using innovated SM processes.

5.1 *Future scope*

In future, additional factors may be identified and prioritized using any other MCDM approach. Barriers and enablers can be identified and investigated in future. Investigation of implementation strategies and enabling technologies can be done in future.

Appendix A

A.1. *Summary of the notations used in this study*

F — Factor
C — Criteria
D — Number of decision-makers
P_{ij} — Aggregation of ratings of ith factor and jth criteria in trapezoidal fuzzy numbers
$\mathcal{U}_{ij}$ — Normalization of ratings of ith factor and jth criteria in trapezoidal fuzzy numbers

$\mathfrak{F}_{ij}$ — Defuzzification of factor ratings
$\mathfrak{F}^{*}_{j}$ — Best value of jth criteria
$\mathfrak{F}^{-}_{j}$ — Worst value of jth criteria
W_{j} — Weight of jth criteria
R_{i} — Regret index of ith factor
$\acute{S}_{i}$ — Utility index of ith factor
Q_{i} — VIKOR index of ith factor

A.2. *Sample calculation to compute three indices: Utility, regret and VIKOR*

For Factor F1,

utility $\acute{S}_{1}$ = {(1.43(1.63 − 1.63))/((1.63 − 1.14)) + (0.71(0.86 − 0.71))/ ((0.86 − 0.39)) + (0.8(0.89 − 0.71))/((0.89 − 0.49)) + (0.86(0.89 − 0.86))/((0.89 − 0.59)) + (0.86(0.86 − 0.86))/ ((0.86 − 0.46))} = 0.68

regret R_{1} = max{(1.43(1.63 − 1.63))/((1.63 − 1.14)), (0.71(0.86 − 0.71))/((0.86 − 0.39)), (0.8(0.89 − 0.71))/((0.89 − 0.49)), (0.86(0.89 − 0.86))/((0.89 − 0.59)), (0.86(0.86 − 0.86))/ ((0.86 − 0.46))} = 0.36

From "$\acute{S}_{i}$" and "R_{i}" values,

$\acute{S}^{*}$ = min (0.68, 2.72, 2.33, 1.53, 1.7, 0.26, 1.97, 0.82, 0.84, 1.6, 1.91, 0.8, 1.58, 1.73, 2.13, 0.84, 1.33, 1.8) = 0.26

$\acute{S}^{-}$ = max (0.68, 2.72, 2.33, 1.53, 1.7, 0.26, 1.97, 0.82, 0.84, 1.6, 1.91, 0.8, 1.58, 1.73, 2.13, 0.84, 1.33, 1.8) = 2.72

R^{*} = min (0.36, 0.8, 0.86, 0.58, 1.17, 0.15, 0.51, 0.58, 0.58, 0.52, 0.86, 0.58, 1.43, 1.17, 1.17, 0.58, 0.58, 1.43) = 0.15

R^{-} = max (0.36, 0.8, 0.86, 0.58, 1.17, 0.15, 0.51, 0.58, 0.58, 0.52, 0.86, 0.58, 1.43, 1.17, 1.17, 0.58, 0.58, 1.43) = 1.43

Then, VIKOR index Q_{1} = (0.5(0.68 − 0.26))/((2.72 − 0.26)) + ((1 − 0.5) (0.36 − 0.15))/((1.43 − 0.15)) = 0.17.

References

Abdel-Basst, M., Mohamed, R., & Elhoseny, M. (2020). A novel framework to evaluate innovation value proposition for smart product–service systems. *Environmental Technology & Innovation*, 20, 101036.

Ali Qalati, S., Li, W., Ahmed, N., Ali Mirani, M., & Khan, A. (2021). Examining the factors affecting SME performance: The mediating role of social media adoption. *Sustainability*, 13(1), 75.

Bogle, I. D. L. (2017). A perspective on smart process manufacturing research challenges for process systems engineers. *Engineering*, 3(2), 161–165.

De Guimarães, J. C. F., Severo, E. A., Campos, D. F., El-Aouar, W. A. and Azevedo, F. L. B. D. (2019). Strategic drivers for product and process innovation: A survey in industrial manufacturing, commerce and services. *Benchmarking: An International Journal*, 27(3), 1159–1187.

Ghobakhloo, M. (2020). Determinants of information and digital technology implementation for smart manufacturing. *International Journal of Production Research*, 58(8), 2384–2405.

Girubha, R. J. & Vinodh, S. (2012). Application of fuzzy VIKOR and environmental impact analysis for material selection of an automotive component. *Materials & Design*, 37, 478–486.

Gumbi, L. & Twinomurinzi, H. (2020). SMME readiness for smart manufacturing (4IR) adoption: A systematic review. In *Conference on e-Business, e-Services and e-Society* (pp. 41–54). Springer, Cham (April 2020).

Horváth, D. & Szabó, R. Z. (2019). Driving forces and barriers of Industry 4.0: Do multinational and small and medium-sized companies have equal opportunities? *Technological Forecasting and Social Change*, 146, 119–132.

Kahle, J. H., Marcon, É., Ghezzi, A. & Frank, A. G. (2020). Smart products value creation in SMEs innovation ecosystems. *Technological Forecasting and Social Change*, 156, 120024.

Larsen, M. S. S. & Lassen, A. H. (2020). Design parameters for smart manufacturing innovation processes. *Procedia CIRP*, 93, 365–370.

Lee, Y. T., Kumaraguru, S., Jain, S., Robinson, S., Helu, M., Hatim, Q. Y., Rachuri, S., Dornfeld, D., Saldana, C. J., & Kumara, S. (2017). A classification scheme for smart manufacturing systems' performance metrics. *Smart and Sustainable Manufacturing Systems*, 1(1), 52.

Lin, T. C., Sheng, M. L., & Jeng Wang, K. (2020). Dynamic capabilities for smart manufacturing transformation by manufacturing enterprises. *Asian Journal of Technology Innovation*, 28(3), 403–426.

Lu, Y., Morris, K. C., & Frechette, S. (2016). Current standards landscape for smart manufacturing systems. National Institute of Standards and Technology, *NISTIR*, 8107, 39.

Malaga, A. & Vinodh, S. (2021a). Benchmarking smart manufacturing drivers using Grey TOPSIS and COPRAS-G approaches. *Benchmarking: An International Journal*. DOI: 10.1108/BIJ-12-2020-0620.

Malaga, A. & Vinodh, S. (2021b). State of art review on smart manufacturing. *International Journal of Business Excellence*. DOI: 10.1504/IJBEX.2020. 10036875.

Mani, Z. & Chouk, I. (2017). Drivers of consumers' resistance to smart products. *Journal of Marketing Management*, 33(1–2), 76–97.

Marzi, G., Dabić, M., Daim, T. & Garces, E. (2017). Product and process innovation in manufacturing firms: A 30-year bibliometric analysis. *Scientometrics*, 113(2), 673–704.

Nimbalkar, S., Supekar, S. D., Meadows, W., Wenning, T., Guo, W. & Cresko, J. (2020). Enhancing operational performance and productivity benefits in breweries through smart manufacturing technologies. *Journal of Advanced Manufacturing and Processing*, 2(4), e10064.

Opricovic, S. & Tzeng, G. H. (2004). Compromise solution by MCDM methods: A comparative analysis of VIKOR and TOPSIS. *European Journal of Operational Research*, 156(2), 445–455.

Sjödin, D. R., Parida, V., Leksell, M., & Petrovic, A. (2018). Smart factory implementation and process innovation. *Research-Technology Management*, 61(5), 22–31.

Travaglioni, M., Ferazzoli, A., Petrillo, A., Cioffi, R., De Felice, F. & Piscitelli, G. (2020). Digital manufacturing challenges through open innovation perspective. *Procedia Manufacturing*, 42, 165–172.

Vinodh, S., Nagaraj, S., & Girubha, J. (2014). Application of Fuzzy VIKOR for selection of rapid prototyping technologies in an agile environment. *Rapid Prototyping Journal*, 20(6), 523–532.

Wankhede V.A. & Vinodh S. (2021). Design strategies enabling industry 4.0. In Chakrabarti A., Poovaiah R., Bokil P. & Kant V. (eds.) *Design for Tomorrow* — Vol. 2. Smart Innovation, Systems and Technologies, Vol. 222. Springer: Singapore. https://doi.org/10.1007/978-981-16-0119-4_64.

Zheng, P., Wang, Z., Chen, C. H. & Khoo, L. P. (2019). A survey of smart product-service systems: Key aspects, challenges and future perspectives. *Advanced Engineering Informatics*, 42, 100973.

Part 2

Artificial Intelligence

Chapter 6

AI-Driven Innovation: Leveraging Big Data Analytics for Innovation

Aniekan Essien

Management Department, Jubilee Building,
University of Sussex Business School, Brighton, UK
a.e.essien@sussex.ac.uk

Technological advancement and data ubiquity have resulted in the rapid and continuous development of data computing and processing tools, hardware, and software, which have impacted all facets of human endeavor. Artificial Intelligence (AI) is increasingly being applied in non-routine tasks that were earlier performed by humans alone. Despite the wave of AI in automation felt in other industries, the innovation sector is yet to feel a widespread adoption of AI in the innovation process. This chapter builds on existing literature to conceptualize the application of innovation analytics — referring to the application of AI-enabled, data-driven insights, algorithms, and visualizations within the innovation process. It is argued that AI has the potential to play a vital role in fostering innovation by driving key aspects within the innovation process. Using a hypothetical example of a tech start-up, we show an example of how infusing AI/big data analytics can serve as key enablers/triggers in the overall innovation process. Further, the chapter explicates the benefits and limitations of using AI in innovation and concludes by providing some implications of applying this technique in the innovation process.

1. Introduction

Data ubiquity is fostering interconnectivity in the world we live in today. Recent reports have shown that the total amount of data generated is forecasted to reach 180 zettabytes (10^{21}) by 2025 (Holst, 2021). To put this in perspective, the world is collectively outputting 2.5 quintillion bytes (10^9) of data daily, with each person generating about 1.7 megabytes of data every second (Reinsel *et al.*, 2017). Evidently, the world today is a collection of data-creating processes and therefore nearly all activities create, copy, or transmit data. For instance, when streaming a song online, data about the time, location, and platform are generated, transmitted, and/or stored somewhere. What about the smartphones and connected devices using the Internet of Things (IoT), whereby home thermostats can be controlled using smartphone applications? Data are being generated (and transmitted/stored) in this instance about the user behavior. If we also consider our fitness/activity wearable devices (smart watches, etc.), data are being generated about how fast we run, heart rates, etc. It is very easy to expand this list to almost every aspect of our daily lives — focusing on interactions with our smartphones. These smartphones act as sensors (same as IoT devices) that measure, collect, and transmit these data, which are typically analyzed to enhance service provision and delivery, shared with other organizations, all resulting in a continuous value chain of data generation processes. Concurrently, technological advancement has resulted in the proliferation of analytical tools, algorithms, and models for analyzing these structured and unstructured/big (and extreme) data — birthing a new field known as big data analytics (BDA), which is a subset of artificial intelligence (AI). It is important to mention that BDA represents a subset of AI that concerns the complex process of examining big data to uncover information (Subramaniyan *et al.*, 2021). Simply put, BDA often refers to the complex process of analyzing big data to extract knowledge in the form of hidden patterns, correlations, etc. (LaValle *et al.*, 2011).

The vastness and relevance of these data have transformed BDA into a vital tool for businesses such that research interest — both scholarly and in practice — in the topic has been consistently on the rise. In the literature, there are many studies that have discussed and analyzed the opportunities brought about by BDA, for instance, Hamilton & Sodeman (2020), Buganza *et al.*, (2015), and George & Lin (2017). Recently, scholars in the innovation research field have considered the strategies that can leverage BDA toward fostering innovation by acting as innovation triggers or indicators within firms (Trabucchi *et al.*, 2018). The past few years have ushered in the "golden age" of digital innovation (Fichman *et al.*, 2014), resulting in a more pervasive and ubiquitous diffusion of digital technologies, which rapidly disrupt the innovation process. Digital innovation refers to innovation processes that rely on digital technologies to offer new products, services, or processes (Martínez-López & Casillas, 2013; Nambisan *et al.*, 2017). Indeed, there are recent scholarly efforts investigating the impact of new and emerging technologies on the innovation process, the output of which is being used as new sources of competitive advantage (Berman, 2012; Hanelt *et al.*, 2021). An example of this digital innovation is data-driven innovation (DDI), which is an innovation project that is birthed by trends or correlations in data (Trabucchi & Buganza, 2019). A relatable application of this data-driven innovation is in the area of decentralized innovation, where AI is applied to build collaborative innovation platforms. It is, in other words, a situation where inventors create the same patent and work together, so there is a link between them. This data-driven approach toward innovation is an aspect of business that can result in tremendous competitive advantage.

In recent times, AI and BDA as research fields have been gaining interest and are being applied in virtually all sectors of human endeavor — engineering (Essien & Giannetti, 2020), science, education (Essien *et al.*, 2021), business, transportation management (Essien *et al.*, 2019, 2020) economics, among others. The research field has seen tremendous growth, so much so that it is almost impossible to effectively track the progression of studies around this topical subject. AI is considered a vital avenue for processing intelligence, listed in Gartner's Top 10 strategic technology trends for 2020 (Gartner, 2019). According to Russell & Norvig (2016), AI involves instilling machines with intelligence that is mainly attributed to humans. More generally, there is a consensus in the academic community that the application of AI/BDA will affect some business activities more than others, and this impact will be greatly dependent on the degree of "creativity" obtainable in the activity. Putting more perspective to this, one can agree that AI/BDA might easily replace humans in repetitive or mundane tasks (e.g., data collection and entry, web crawling, and data cleaning) more than the rather creative activities. Therefore, the greater the level or creativity, the more difficult it becomes for AI to add value. This has become a fascinating dilemma for innovation managers: How can AI/BDA support innovation when highly creative activities cannot be automated? In this chapter, a discussion using practical examples is presented as to how BDA — specifically AI techniques — can act as a trigger in the innovation process, as well as describing using hypothetical examples how these AI- and data-driven techniques can be applied within the innovation process. To achieve this, a conceptual research approach is applied, using a conceptual case study of a tech start-up firm that significantly relies on digital technology, thereby exposing the potential of infusing data analytics to serve as a key enabler/trigger in the overall innovation process.

2. Literature Review: The Role of AI/Big Data Analytics in Innovation

The core of innovation encompasses an iterative process that mainly aims at creating new products, processes, or services by using new or existing knowledge (Kusiak, 2009). On the other hand, a data-driven innovation regime corresponds to the manipulation of any data within the innovation process for the purpose of value creation (Stone & Wang, 2014). Consequently, data plays a critical role in the innovation process, emerging a new trend where data and analytics are driving innovation. This is achieved by the proliferation of data-driven goods and services, which are enabling better planning, marketing, and operations that are informed by data-driven intelligence and is being applied across a broad range of industrial sectors. Taking an economic and retrospective perspective, data can be comparable in terms of returns and potential as oil serves in the infrastructure by comprising many actors that are exploited for various competing or complementary approaches. In general, the increasing demand for data is accelerated by the downstream activities that now greatly rely on data as input, for instance, e-commerce, global supply chain management, and marketing. In other words, data offer a potentially significant return in terms of scale and scope.

Big data as a research topic/field is now gaining a lot of popularity. There have been a lot of definitions of big data over time, such that it is rather difficult to keep up with a consensus definition of the term. However, some articles do share similarities in the key descriptive for this terminology (Brown *et al.*, 2011; De Mauro *et al.*, 2016; Eassom, 2015; Gandomi & Haider, 2015). Popular among these is the concept of the 3Vs model (Anshari & Lim, 2016; McAfee *et al.*, 2012). These models consider the role of (i) volume, relating to the huge amount of data, (ii) the variety of the data, relating to the fact that the data are obtainable from various sources, and (iii) the velocity of the data, meaning that the data are inputted in a continuous stream. Later on, some other Vs were added (Gandomi & Haider, 2015) to incorporate the concepts of veracity (i.e., the use of reliable data and confident interpretations), variability (i.e., the management and interpretation of continuous data stream and its changes), and value (relating to the exploitation of the value embedded in the data).

In recent times, there have been trends considered relevant in the enhancement and enablement of big data in driving various aspects of human endeavor (Dumbill, 2013). These have resulted in many benefits, including the diffusion of smart devices (e.g., phones, tablets, and wearable devices) to social networking platforms, which have assumed key roles acting as data generators (*New York Times*, 2012). However, the proliferation and diffusion of cloud computing has provided an option for the access and storage of data, which has significantly boosted its relevance (Marston *et al.*, 2011) alongside the cost reduction implied in computational power due to the advancement in technology (Vajjhala & Ramollari, 2016). Further, there are other significant trends that are rising and expanding the adoption and diffusion of big data, such as the IoT, AI, and machine learning (ML). Individually, these concepts serve as paradigms that can be considered as the outcome of a convergence process of various visions relating to technology, which integrates objects, connectivity, and adopt a semantic approach (Atzori *et al.*, 2010). This concept is typically defined *as a global infrastructure for the information society, enabling advanced services by interconnecting (physical and virtual) things based on existing and evolving interoperable information and communication technologies.* On the other hand, humans and their behavior or opinions are considered part of an evolving ecosystem, which enables even more critical (potential) criticalities with respect to big data and analytics. The peculiarity and potential of integrating the physical and infrastructural technological components/elements and people in data generation have been discussed in the concept of *Internet of People* (Conti *et al.*, 2017), which is defined as a complex socio-technical system involving a situation where humans interact with their devices representing the key nodes of a network that creates valuable and exploitable data.

On the extreme side of this spectrum, AI is rapidly becoming a popular topic in business and its application has been explored in many disciplines in both industry and academia (Chui *et al.*, 2018; Kietzmann & Pitt, 2020). A coherent conclusion is that AI will surely impact more businesses than others, depending on the degree of inherent creativity relating to the activity. AI is all about instilling machines with a kind of intelligence that is attributable to humans (Russell & Norvig, 2016). According to Kakatkar *et al.* (2020), AI can play a significant role in creating and enabling innovation across the data-driven innovation process. AI and data-driven innovations are subliminally associated with the value chain model, which is sometimes referred to as a "virtual value chain", indicating the extent to which the data will be of interest with respect to how it will be gathered, organized, selected, transformed into products or services, and distributed (Piccoli, 2007; Rayport & Sviokla, 1996). AI/BDA is at the intersection of driving and delivering data-driven innovation — accentuated by technological advancement. At the organizational level, at least two categories of strategic initiatives could result from big data-driven innovation and its underlying big data value chain. The prior group of advantages has the main aim of ensuring the full availability of information pertaining to aspects of organizational processes and services to enable improvements. In general, big data is capable of instrumenting organizational operations to inform or drive requisite change in these organizations. Furthermore, as argued in Piccoli (2007), there are also the external facing initiatives, involving the inherent exploitation of customer data, including but not limited to consumer and customer logs, transaction records, consumer-generated data, and other customer-generated contents applied toward fostering direct and implied marketing campaigns, targeted and personalized recommendations, which result in increased sale and customer satisfaction. A typical example of this exploitation of data-driven innovation can be seen in Netflix's collaborative filtering algorithm to predict user movie ratings (Chen *et al.*, 2012). Another example is how Google uses users' search behavior to target advertising.

Recent scholarly efforts have shown the impact of big data on different kind of innovations, ranging from process to product to paradigm to positional innovation, incorporating an architecture to a modular base (Caputo *et al.*, 2016; George & Lin, 2017). On the one hand, AI and big data analytics can help companies improve the handling of extant resources. So, data can be used to drive intelligence and analytics within the developmental space in fostering a so-called analytics-driven innovation based on data (Tempini, 2017). On the other hand, AI and big data analytics can be applied to drive personalization or recommendation services in service organizations/operations, leveraging data as an archetype (Ng & Wakenshaw, 2017). Conversely, it may be possible to integrate AI/BDA toward an extension of the capabilities of a firm beyond its boundaries, as observed in Pellegrini (2017). This way, AI/BDA can drive an open innovation strategy that can lead to new business opportunities in potential areas of immediate gain (Del Vecchio *et al.*, 2018), which can be applied toward the pursuit and implementation of business model innovation toward the enlargement of the company domain (Trabucchi *et al.*, 2018). On the contrary, there has been recent interest from scholars in delving deeper into the key enablers within the organizations that can serve as facilitators in this data-driven innovation, for instance, data management and customer or consumer centricity. (Troilo *et al.*, 2017).

Recent research in the topical area of AI and big data-driven innovation has shown how these emerging and existent technologies can be applied toward challenging traditional two-sided markets based on the advertising mechanisms, which can be seen as a non-transaction two-sided market, given that the manifesto is not able to permit a direct transaction between the competing sides. A typical example of this is where some marketing businesses leverage a two-sided configuration, especially when it is considered that a single side (the advertisers) is constantly looking in a forward direction to reach the first side (the customers or clients) via a Client-as-a-Target standpoint to win over their attention (Filistrucchi *et al.*, 2014; Trabucchi *et al.*, 2018). In this format, the data obtained on this digital platform (e.g., a mobile-based app) can reverse the fundamental relationship existing between the two sides, given that the technological platform might leverage one of the abovementioned strategies toward engaging a particular group of customers on the reverse side that evidences some interest in the gathered data. From the foregoing, it becomes important to highlight the recent developments in the various ways in which innovation can be pursued using AI/BDA. For instance, in Chang (2018), there is an argument for the need for further research on how AI and big data analytics can provide motivation to promote innovation, especially while recent dynamic and constant technological evolution has been driving the chance to assemble specific data in well-timed and cheaper ways (Gandomi & Haider, 2015). Considering the same viewpoint, Rindfleisch *et al.* (2017) present an articulate discourse on how AI and big data analytics can serve a dual function of allowing firms to exploit their inherent value (innovation from data) as well as enabling a specific approach toward product innovation, further driven by the opportunity to obtain and analyze data (i.e., innovation as data). This chapter aims to present a discussion using practical examples of how AI/BDA can act as a trigger in the innovation process, as well as describe using examples how these AI- and data-driven techniques can be applied within the innovation process.

3. Innovation Analytics: Using AI to Drive Innovation

When one considers the broad term of "innovation", it is clear from recent times that there is a widespread increase of studies researching the impact of technological innovations on the fates of companies and the dynamics of industries. In other words, the ability of a particular company to capture value from new/emerging technologies, markets, and business models is a key factor in deciding competitive edge (i.e., which company will be a leader in its industry or will be disrupted/out of the market). This is also the position of innovation scholars, such as Christensen (2013), Christensen & Raynor (2013), and Tidd & Bessant (2020). Innovation is broadly classified as incremental or radical, with the former referring to a situation where innovation takes place using a set of rules that are clearly understood and having players "innovating" by doing what they have been doing, but better (Christensen, 2013). In the literature, this incremental innovation is one that produces incremental or small changes by introducing new features to existing technologies, for instance, companies gradually improving while building new capabilities or a company upgrading its enterprise resource planning (ERP) (Christensen & Raynor, 2013). On the other hand, it is sometimes the case that an industry is significantly/completely disrupted, changing "the rules of the game", not only opening up new opportunities but also challenging existing players to reframe what they are doing in light of the new conditions (Leifer *et al.*, 2000; O'Connor *et al.*, 2008). According to The Innovator's Dilemma (Christensen, 2013), the main aim of incremental innovation is to satisfy the needs of customers, improving product performance using feedback obtained from the customers, reducing defects, and making a product or service faster and more powerful. On the other hand, disruptive innovation satisfies customer future needs and may provide lower performance in some key features but creates some unique features valued by market. Therefore, in this section, there is a characterization of innovation analytics in both radical and incremental innovation.

3.1. *Innovation analytics for radical innovation*

The proliferation of AI/BDA has seen computers and computing devices (e.g., IoT and wearables) play a markedly dominant role in the innovation process, contrary to the conventional servile role adopted for handling tasks that the innovation manager might consider monotonous or arduous to manually perform. AI — and technology broadly speaking — has been and will always be a key enabler for disruptive innovation that leads to game-changing products and services able to serve consumers at the low end as well as migrating to the mainstream market. In fact, AI/BDA is at the heart of the disruptive innovation that is currently being witnessed — from Google/Apple Maps to the virtual assistants (i.e., Google Alexa or Apple Siri). In the organizational perspective, a new business model has emerged in the AI-enabled sharing economy where individuals can share their own assets with others as services to make better use of these assets and to generate profit (e.g., Uber, Airbnb, etc.).

Typical examples of these tasks can encompass routine data processing, results replication and/or broadcasting, storing data, and simple database actions (e.g., backup and health check routines). AI has the potential to rapidly expand this role in the innovation process by taking computers to act as co-innovators and partners, further accentuating the creative values and strengths in humans. AI has the capability to perform deep analysis on big data (i.e., big data analytics) via uncovering patterns in hierarchical big structured and unstructured datasets, spotting anomalies, etc. By adopting these inherent capabilities in AI-driven analytics, data scientists — using the broad range of algorithms, tools, and applications — can provide innovation managers relevant assistance toward unlocking the potential value obtainable by integrating AI and innovation projects.

3.2. *Innovation analytics for incremental/sustaining innovation*

AI/BDA can also be applied toward sustained or incremental innovation regimes. In this way, the application of AI/BDA can be toward satisfying the needs of customers, improving product performance using feedback obtained from the customers, reducing defects, and making a product or service faster and more powerful (Christensen, 2013). Kakatkar *et al.*, (2020) posit that AI can impact at least four main drivers in innovation analytics: (i) specifying objectives, which can be mapped to AI-enabled analyses involving descriptive, predictive, prescriptive, and diagnostic analyses; (ii) articulating data collection and preparation; (iii) data modeling; and (iv) capturing value in the innovation process. Consequently, AI can play a significant role in the innovation process, ranging from problem exploration to solution selection and implementation. In short, AI has the potential to drive innovation analytics, and this can be realized by its application in innovation processes. Therefore, it is possible to articulate some implications on how AI can drive innovation by how the innovation managers consider the possibility of leveraging technology in the innovation management process. Kakatkar *et al.* (2020) noted that AI can result in considerable value from the availability of big data by allowing innovation teams to leverage scores of big data available due to data ubiquity for the purpose of executing insightful analyses that are highly scalable and replicable. However, there is still an opportunity to enhance the plethora of algorithms designed for parsing both structured and unstructured data (Kakatkar & Spann, 2019; Wedel & Kannan, 2016). These tools have the capability to cover specific aspects in the innovation process, which can be beneficial in sentiment analysis, recommender systems, and content coding.

Furthermore, AI can inspire innovation managers to collaborate with data scientists to delegate tasks having higher creative complexity to the computers, suggesting that AI can help validate creative insights and minimize creativity in the specific aspects. AI can also enable the key stakeholders in the innovation process to provide better answers to existing questions and permit the proliferation and application of AI algorithms and models to uncovering hidden patterns between complex variables and deriving new hypotheses about the innovation process (Puranam *et al.*, 2018). Applying AI in innovation does have its own demerits. For instance, there is an inherent challenge with conventional innovation teams which might lack the technical know-how for building and using AI models, necessitating the need for collaboration with data scientists, ideally making them core team members from the start. Besides, current technological limitations result in short-term restrictions in the output of the AI models. For instance, the output from these models may not be as contextually toned as human or manually curated analyses.

4. AI-driven Innovation: Integrating AI/BDA to Drive Innovation

Linda Hill in the book, *Collective Genius*, references the concept of creative resolution (Hill *et al.*, 2014), asserting the (now known) reality that the biggest discoveries and inventions (and innovations) result from the connection of ideas from diverse and seemingly disparate approaches and people. Even more, the empirical study in Østergaard, Timmermans, & Kristinsson (2011) investigating the relation between diversity in the knowledge base and the performance of firms showed that there is a positive relation between diversity in education and gender on the likelihood of introducing an innovation, as echoed in more recent articles (Attah-Boakye *et al.*, 2020; Ferrucci, 2020; Zouaghi *et al.*, 2020). However, the argument presented in this chapter takes a more divergent approach toward disparity and diversity — referring to diversity of data and computing. In other words, there is an argument for the fusion and integration of data and design to drive — product and process — innovation.

From the points made in the preceding sections of this chapter, we can easily visualize the potential inherent in big data (consider the 5Vs), with data typically serving as critical inputs to many business and organizational processes. Further, the proliferation of technology and computing resulting in advances in AI/BDA has further contributed toward enhancing the capability and feasibility with which decision-makers can leverage support (and confidence) from data to make better informed decisions — ranging from hiring to product development, customer engagement, and market segmentation. The technological giants (a.k.a. GAFAT — Google, Amazon, Facebook, Apple, and recently Tesla) have continuously demonstrated the immense economic and customer-centric benefits realizable by integrating analytics within their design thinking process. This extends from harnessing qualitative insights to a relentless focus on the customer needs, with technology at the heart of it all. However, despite these benefits, organizations still somehow appear short in reaping the rewards of innovation analytics and, therefore, miss out on the benefits that are obtainable thereof. Indeed, in a McKinsey report (Akshay & Williams, 2019), it was reported that on average, there is a 10–30% performance improvement among companies that integrate some sort of data analytics within the innovation design process (for product innovation). The same report, which surveyed the design practices of 300 companies, showed a strong correlation between the top financial performers and an innovation process that infuses data in its design. The authors recommended the assembly of a robust "squad" that includes both data experts (e.g., data scientists, engineers, and administrators) and research experts, innovation managers, etc., all of whom should be on the same team. This closely tallies with the findings from studies relating to diversity and innovation (Attah-Boakye *et al.*, 2020; Ferrucci, 2020; Zouaghi *et al.*, 2020).

In this chapter, to demonstrate the possibility and potential involved in data-driven innovation (or innovation analytics), we discuss a hypothetical case study involving a fictitious tech start-up, *FSP*. It is important to mention at this point that instead of presenting a deep dive into the specifics of how the AI-driven innovation process can work, the chapter overall aims to expose the potential of infusing data analytics to serve as a key enabler/trigger in the overall innovation process.

Let us consider *FSP*, which is a fictitious manufacturer of integrated circuits. As can be immediately envisioned, a company of this nature will mainly have other manufacturers (e.g., automobile, mobile, and computer manufacturers) as its direct customers. However, despite predominantly trading as a B2B enterprise, *FSP* realizes the need to ensure that its customers are served excellently. As is known within the marketing space, a key challenge with maintaining B2B relationships (and other relationships in general) is the presence of a customer base with a broad range of specific requirements and needs. Therefore, the customer base of *FSP* might comprise consumers (at all levels), professionals, other technological giants, and tech enthusiasts who may consume these products.

With this knowledge in mind, *FSP* can adopt an AI- and data-driven approach toward solving the problem of keeping the diverse customer base very much satisfied, thereby shaping their views and enhancing their experience. The first step in this process is the assembly of a team capable of achieving this. According to Akshay & Williams (2019), this is the most effective and important step in innovation analytics, as with these squads, the firm can achieve a more iterative inquiry process and take advantage of the best of both worlds — a concept that Linda Hill refers to as *creative resolution* (Hill *et al.*, 2014). Therefore, the management of *FSP* proceeds to set up a diverse and cross-functional team comprising data scientists, data engineers, and R&D researchers. The main aim of this team is to analyze the data (transactional, purchase, demographic, etc.) from the end users with the motive of identifying potential lead users, which can be used in understanding the user needs and problems. The immense benefit of infusing AI in this analysis process cannot be overstated. In short, the only way this type of analysis could have been done without AI is to manually survey (e.g., questionnaires or qualitative surveys) the customer base or by adopting analyses of user-generated content obtainable from social media platforms, etc., to derive these potential insights that can be obtained using AI. Therefore, adopting an AI-driven approach can benefit the company by fostering the identification of their lead customer base including their purchase behavior and preferences, as well as uncovering any key problem areas that can be identifiable by this analysis.

A second important step in this process of innovation analytics relates to data collection and preparation. According to Akshay & Williams (2019), the facilitation and development of cross-functional knowledge can be enhanced by training the data scientists to understand (and appreciate) how design research is done, the value it brings, and, conversely, enabling designers to gain an inside view of the work of data scientists and better realize how data can be harnessed. In this data collection and preparation process, the data scientists and data engineers sieve through a mountain heap of data, scanning through IP addresses, social media platforms, click history, purchase and transaction records, etc., to uncover rich insight from the dataset. This can be achieved by developing AI-driven (deep-learning) models, which are advantageous in automatic feature extraction (Essien & Giannetti, 2020). The model can be trained to classify or sieve through the user-generated datasets to filter out relevant posts.

In identifying the lead users, the data scientists can adopt an unsupervised approach toward clustering and segmenting the resulting dataset. The rationale for adopting an unsupervised approach is its benefit in analyzing large datasets that typically lack ready labels that have been previously classified. Besides, another benefit of adopting unsupervised learning is its ability to learn and produce classifications without supervision (Nanduri & Sherry, 2016). With respect to the features that can be used in this unsupervised model, some key variables are account popularity (e.g., based on number of posts, likes, or comments) and the transaction volume encountered in the process. In terms of value capturing, a simple sentiment analysis approach can be adopted in analyzing the degree of satisfaction of the customer, computing a sentiment score.

From the contrived and hypothetical case study presented, it can be seen that *FSP* can realize knowledge and understanding about its customers, delving deeper into the level of being able to offer customized solutions and offerings to the potential lead customers, in addition to uncovering problem or challenge areas in the existing relationships. By adopting this AI-driven innovation analytics, the efficiency of the teams is greatly enhanced and a higher success rate is guaranteed at a significantly lower (financial and time) project cost. However, it is important to consider this illustration with some limitations. First, there is a challenge relating to the authenticity and verification of the input data. Since social media does not mandate user verification, there is the possibility of some fake or bogus accounts, which may alter the accuracy of the model that has been developed and implemented.

5. Conclusion

The wave of technological advancement observed in the past decade has resulted in the proliferation of data and computing, which has impacted nearly all facets of human endeavor. Synchronously, this technological advancement has resulted in the rapid and continuous development of analytical tools, algorithms, and models for analyzing the huge amounts of data available today. Therefore, technological giants have adopted these techniques — referred to as big data analytics (BDA) — for analyzing big data to extract knowledge in the form of hidden patterns.

In recent times, the concept of digital innovation referring to innovation processes that rely on digital technologies to offer new products, services, or processes has fast gained popularity, with data-driven innovation (DDI) taking the spotlight in this chapter (Trabucchi & Buganza, 2019). This data-centric innovation is an aspect of business that can result in tremendous competitive advantage. This chapter has shown the potential that can be obtained by infusing data analytics and AI in driving innovation, using a hypothetical example to showcase how AI/BDA can act as a trigger in the innovation process. By applying a conceptual approach of a conceptual case study involving a tech start-up firm, we show an example of how infusing data analytics can serve as a key enabler/trigger in the overall innovation process.

Although this chapter presented an extremely hypothetical and simplified example, the underlying principles and potential benefits remain the same. By applying the principles set out in existing scholarly efforts detailing the implementation of innovation analytics, firms can leverage these tools and techniques to realize greater value obtainable in fusing data and innovation. It is important not to underestimate the challenge ensuing with adopting and implementing the concept presented in this chapter. For a firm to adopt and implement this AI-driven analytics, there must be a deliberate and consistent drive toward assembling a capable team (comprising both R&D and data engineers/scientists), which is the most effective and important step in innovation analytics, as with these squads, the firm can achieve a more iterative inquiry process and take advantage of the best of both worlds. Achieving this requires the firm to build capabilities and competencies around this team/squad, which is the main driver and determinant of the success of such an approach. The other steps to capturing value from innovation analytics rely on the abilities and technical nous of this assembled team. Therefore, the greater the (careful and articulate) integration of AI, big data analytics, and innovation, the greater the potential financial benefit realized by the firm.

References

Akshay, C. & Williams, S. (2019). Fusing data and design to supercharge innovation--in products and processes | McKinsey. https://www.mckinsey.com/business-functions/mckinsey-analytics/our-insights/fusing-data-and-design-to-supercharge-innovation-in-products-and-processes.

Anshari, M. & Lim, S. A. (2016). Customer relationship management with big data enabled in banking sector. *Journal of Scientific Research and Development*, 3(4), 3.

Attah-Boakye, R., Adams, K., Kimani, D., & Ullah, S. (2020). The impact of board gender diversity and national culture on corporate innovation: A multi-country analysis of multinational corporations operating in emerging economies. *Technological Forecasting and Social Change*, 161, 120247.

Atzori, L., Iera, A., & Morabito, G. (2010). The internet of things: A survey. *Computer Networks*, 54(15), 2787–2805.

Berman, S. J. (2012). Digital transformation: Opportunities to create new business models. *Strategy & Leadership*, 40, 16–24.

Brown, B., Chui, M., & Manyika, J. (2011). Are you ready for the era of "Big Data." *McKinsey Quarterly*, 4(1), 24–35.

Buganza, T., Dell'Era, C., Pellizzoni, E., Trabucchi, D., & Verganti, R. (2015). Unveiling the potentialities provided by new technologies: A process to pursue technology epiphanies in the smartphone app industry. *Creativity and Innovation Management*, 24(3), 391–414.

Caputo, A., Marzi, G., & Pellegrini, M. M. (2016). The internet of things in manufacturing innovation processes: Development and application of a conceptual framework. *Business Process Management Journal*, 22(2), 383–402.

Chang, V. (2018). A proposed social network analysis platform for big data analytics. *Technological Forecasting and Social Change*, 130, 57–68.

Chen, H., Chiang, R. H. L., & Storey, V. C. (2012). Business intelligence and analytics: From big data to big impact. *MIS Quarterly*, 36(4), 1165–1188.

Christensen, C. M. (2013). *The Innovator's Dilemma: When New Technologies Cause Great Firms to Fail.* Brighton, Massachusetts: Harvard Business Review Press.

Christensen, C., & Raynor, M. (2013). *The Innovator's Solution: Creating and Sustaining Successful Growth.* Brighton, Massachusetts: Harvard Business Review Press.

Chui, M., Manyika, J., Miremadi, M., Henke, N., Chung, R., Nel, P., & Malhotra, S. (2018). *Notes from the AI Frontier: Insights from Hundreds of Use Cases.* New York: McKinsey Global Institute.

Conti, M., Passarella, A., & Das, S. K. (2017). The Internet of People (IoP): A new wave in pervasive mobile computing. *Pervasive and Mobile Computing*, 41, 1–27.

De Mauro, A., Greco, M., & Grimaldi, M. (2016). A formal definition of Big Data based on its essential features. *Library Review*, 65(3), 122–135.

Del Vecchio, P., Di Minin, A., Petruzzelli, A. M., Panniello, U., & Pirri, S. (2018). Big data for open innovation in SMEs and large corporations: Trends, opportunities, and challenges. *Creativity and Innovation Management*, 27(1), 6–22.

Dumbill, E. (2013). *Making Sense of Big Data.* New Rochelle, NY, USA: Mary Ann Liebert, Inc.

Eassom, S. (2015). IBM Watson for education. *IBM Insights on Business*, 1.

Essien, A., Chukwukelu, G., & Essien, V. (2021). Opportunities and challenges of adopting artificial intelligence for learning and teaching in higher education. In *Fostering Communication and Learning with Underutilized Technologies in Higher Education* (pp. 67–78). UK: IGI Global.

Essien, A. & Giannetti, C. (2020). A deep learning model for Smart Manufacturing using Convolutional LSTM Neural Network Autoencoders. *IEEE Transactions on Industrial Informatics*, 24, 1345–1368.

Essien, A., Petrounias, I., Sampaio, P., & Sampaio, S. (2019). Deep-PRESIMM: Integrating deep learning with microsimulation for traffic prediction. In *Conference Proceedings — IEEE International Conference on Systems,*

Man and Cybernetics, *2019-October*. https://doi.org/10.1109/SMC.2019. 8914604.

Essien, A., Petrounias, I., Sampaio, P., & Sampaio, S. (2020). A deep-learning model for urban traffic flow prediction with traffic events mined from twitter. *World Wide Web*, 24, 1345–1368.

Ferrucci, E. (2020). Migration, innovation and technological diversion: German patenting after the collapse of the Soviet Union. *Research Policy*, 49(9), 104057.

Fichman, R. G., Dos Santos, B. L., & Zheng, Z. (2014). Digital innovation as a fundamental and powerful concept in the information systems curriculum. *MIS Quarterly*, 38(2), 329-A15.

Filistrucchi, L., Geradin, D., Van Damme, E., & Affeldt, P. (2014). Market definition in two-sided markets: Theory and practice. *Journal of Competition Law & Economics*, 10(2), 293–339.

Gandomi, A. & Haider, M. (2015). Beyond the hype: Big Data concepts, methods, and analytics. *International Journal of Information Management*, 35(2), 137–144.

Gartner. (2019). Gartner top 10 strategic technology trends for 2020 — Smarter with Gartner. https://www.gartner.com/smarterwithgartner/gartner-top-10-strategic-technology-trends-for-2020/.

George, G. & Lin, Y. (2017). Analytics, innovation, and organizational adaptation. *Innovation*, 19(1), 16–22.

Hamilton, R. H. & Sodeman, W. A. (2020). The questions we ask: Opportunities and challenges for using big data analytics to strategically manage human capital resources. *Business Horizons*, 63(1), 85–95.

Hanelt, A., Firk, S., Hildebrandt, B., & Kolbe, L. M. (2021). Digital M&A, digital innovation, and firm performance: An empirical investigation. *European Journal of Information Systems*, 30(1), 3–26.

Hill, L. A., Brandeau, G., Truelove, E., & Lineback, K. (2014). *Collective Genius: The Art and Practice of Leading Innovation*. Brighton, MA: Harvard Business Review Press.

Holst, A. (2021). Total data volume worldwide 2010–2025 | Statista. https://www.statista.com/statistics/871513/worldwide-data-created/.

Kakatkar, C., Bilgram, V., & Füller, J. (2020). Innovation analytics: Leveraging artificial intelligence in the innovation process. *Business Horizons*, 63(2), 171–181.

Kakatkar, C. & Spann, M. (2019). Marketing analytics using anonymized and fragmented tracking data. *International Journal of Research in Marketing*, 36(1), 117–136.

Kietzmann, J. & Pitt, L. F. (2020). Artificial intelligence and machine learning: What managers need to know. *Business Horizons*, 63(2), 131–133.

Kusiak, A. (2009). Innovation: A data-driven approach. *International Journal of Production Economics*, 122(1), 440–448.

LaValle, S., Lesser, E., Shockley, R., Hopkins, M. S., & Kruschwitz, N. (2011). Big data, analytics and the path from insights to value. *MIT Sloan Management Review*, 52(2), 21–32.

Leifer, R., McDermott, C. M., O'connor, G. C., Peters, L. S., Rice, M. P., & Veryzer Jr., R. W. (2000). *Radical Innovation: How Mature Companies Can Outsmart Upstarts*. Brighton, Massachusetts: Harvard Business Press.

Marston, S., Li, Z., Bandyopadhyay, S., Zhang, J., & Ghalsasi, A. (2011). Cloud computing — The business perspective. *Decision Support Systems*, 51(1), 176–189.

Martínez-López, F. J. & Casillas, J. (2013). Artificial intelligence-based systems applied in industrial marketing: An historical overview, current and future insights. *Industrial Marketing Management*, 42(4), 489–495.

McAfee, A., Brynjolfsson, E., Davenport, T. H., Patil, D. J., & Barton, D. (2012). Big Data: The management revolution. *Harvard Business Review*, 90(10), 60–68.

Nambisan, S., Lyytinen, K., Majchrzak, A., & Song, M. (2017). Digital innovation management: Reinventing innovation management research in a digital world. *Mis Quarterly*, 41(1), 223–238. DOI: 10.25300/MISQ/2017/41:1.03.

Nanduri, A. & Sherry, L. (2016). Anomaly detection in aircraft data using Recurrent Neural Networks (RNN). *ICNS 2016: Securing an Integrated CNS System to Meet Future Challenges*, 5C2-1–5C2-8. https://doi.org/10.1109/ICNSURV.2016.7486356.

New York Times. (2012). Big Data Impact.

Ng, I. C. L. & Wakenshaw, S. Y. L. (2017). The Internet-of-Things: Review and research directions. *International Journal of Research in Marketing*, 34(1), 3–21.

O'Connor, G. C., Paulson, A. S., & DeMartino, R. (2008). Organizational approaches to building a radical innovation dynamic capability. *International Journal of Technology Management*, 44(1–2), 179–204.

Østergaard, C. R., Timmermans, B., & Kristinsson, K. (2011). Does a different view create something new? The effect of employee diversity on innovation. *Research Policy*, 40(3), 500–509.

Pellegrini, T. (2017). Semantic metadata in the publishing industry — Technological achievements and economic implications. *Electronic Markets*, 27(1), 9–20.

Piccoli, G. (2007). *Information Systems for Managers: Texts and Cases*. New York: Wiley Publishing.

Puranam, P., Shrestha, Y. R., He, F., & von Krogh, G. (2018). *Algorithmic Induction through Machine Learning: Opportunities for Management and Organization Research*. France: INSEAD.

Rayport, J. F. & Sviokla, J. J. (1996). Exploiting the virtual value chain. *McKinsey Quarterly*, 1, 20–36.

Reinsel, D., Gantz, J., & Rydning, J. (2017). Data age 2025: The evolution of data to life-critical don't focus on big data; focus on the data that's big sponsored by seagate the evolution of data to life-critical don't focus on Big Data; Focus on the data that's big. www.idc.com.

Rindfleisch, A., O'Hern, M., & Sachdev, V. (2017). The digital revolution, 3D printing, and innovation as data. *Journal of Product Innovation Management*, 34(5), 681–690.

Russell, S. & Norvig, P. (2016). Artificial intelligence: A modern approach. http://thuvien.thanglong.edu.vn:8081/dspace/handle/DHTL_123456789/4010.

Stone, D. & Wang, R. (2014). Deciding with data — How data-driven innovation is fuelling Australia's economic growth. *PricewaterhouseCoopers (PwC), Melbourne*.

Subramaniyan, M., Skoogh, A., Bokrantz, J., Sheikh, M. A., Thürer, M., & Chang, Q. (2021). Artificial intelligence for throughput bottleneck analysis– State-of-the-art and future directions. *Journal of Manufacturing Systems*, 60, 734–751.

Tempini, N. (2017). Till data do us part: Understanding data-based value creation in data-intensive infrastructures. *Information and Organization*, 27(4), 191–210.

Tidd, J. & Bessant, J. R. (2020). *Managing Innovation: Integrating Technological, Market and Organizational Change*. Hoboken, New Jersey: John Wiley & Sons.

Trabucchi, D. & Buganza, T. (2019). Data-driven innovation: Switching the per-spective on Big Data. *European Journal of Innovation Management*, 22(1), 23–40. DOI: 10.1108/EJIM-01-2018-0017.

Trabucchi, D., Buganza, T., Dell'Era, C., & Pellizzoni, E. (2018). Exploring the inbound and outbound strategies enabled by user generated big data: Evidence from leading smartphone applications. *Creativity and Innovation Management*, 27(1), 42–55.

Troilo, G., De Luca, L. M., & Guenzi, P. (2017). Linking data-rich environments with service innovation in incumbent firms: A conceptual framework and research propositions. *Journal of Product Innovation Management*, 34(5), 617–639.

Vajjhala, N. R. & Ramollari, E. (2016). Big data using cloud computing-opportunities for small and medium-sized enterprises. *European Journal of Economics and Business Studies*, 2(1), 129–137.

Wedel, M. & Kannan, P. K. (2016). Marketing analytics for data-rich environ-ments. *Journal of Marketing*, 80(6), 97–121.

Zouaghi, F., Garcia-Marco, T., & Martinez, M. G. (2020). The link between R&D team diversity and innovative performance: A mediated moderation model. *Technological Forecasting and Social Change*, 161, 120325.

Chapter 7

A Survey of IIoT and AI-Enabled Manufacturing Systems: Use Case Perspective

N. R. Srinivasa Raghavan[*,†] and Sandeep Dulluri[†,§]

*Tarxya Limited, Bangalore, India and
Manchester University, Manchester, UK*
†*Abu Dhabi National Oil Company (ADNOC), Abu Dhabi, UAE*
‡*raghavan@tarxya.com*
§*dulluri@outlook.com*

This chapter surveys the Industrial Internet of Things (IIoT) and artificial intelligence-enabled (AI-enabled) manufacturing systems from a use case perspective. The authors propose a human system ideology of sensory organs and mind to conceptualize the IIoT and AI-enabled manufacturing systems in a novel way. Emphasis is placed on surveying recent scientific articles that provide illustrations and practical applications. Several important problems are captured for interested researchers as directions for future research.

1. Introduction

The manufacturing industry is of pivotal importance, placed next to the agriculture sector. The manufacturing industry has played a key role in nations' development and uplifting their economy. The deep fusion of manufacturing technology with information communication technology, intelligent systems, and product-related expertise has disrupted the manufacturing sector. This amalgamation of disruptors forms an "Industrial Internet of Things" (IIoT) ecosystem.

1.1. *Background and context*

The concept of IIoT stems from the Industry 4.0 that originated in Germany, while the Industrial Internet traces its roots to the United States. From an industrial viewpoint, General Electric (GE) coined the term "Industrial Internet", and Cisco used the term "Internet of Everything" (Gilchrist, 2016). There is a significant overlap in both these concepts, especially from the end user application point of view and hence in the context of this chapter.

In 2012, the GE company proposed the concept of "Industrial Internet", which can connect intelligent equipment, people, and data and analyze such data in an intelligent manner to enable smarter decision-making by humans and machines. There is an evolving need as well as interest in the development of an open and interoperable environment that can be used for Internet Supervisory control and data acquisition (SCADA), Internet-based monitoring, and industrial control systems (Jaloudi, 2019).

GE conceptualized three major components of the "Industrial Internet":

- intelligent equipment,
- intelligent systems,
- intelligent decision-making.

They developed the PREDIX platform with a maximizing potential of the "Industrial Internet" through the holistic integration of the three components — intelligent equipment, intelligent systems, and intelligent decisions — with machines, equipment sets, facilities, and system networks (Gilchrist, 2016). GE pioneered the concept of Digital Twins, i.e., an emulated representation of physical systems with corresponding software modules that are intelligent and able to sense, plan, and execute. Digital Twins in the context of intelligent manufacturing are characterized by autonomy and self-optimization, which proposes new demands such as learning and cognitive capacities for manufacturing cells (Zhou *et al.*,

2020). Connectivity and Integration standards are the key enablers for IIoT in smart manufacturing (Lu *et al.*, 2020). The growth in industrial development due to IIoT is estimated to be over 150 billion dollars in 2020 and is forecasted to continue an accelerated growth path (Gilchrist, 2016).

1.2. *Top 10 disruptors for manufacturing systems*

Gilchrist (2016) mentions that the top 10 disruptors in Industry 4.0 are as follows:

(1) AI & ML
(2) Smart phones and/or mobile devices
(3) Advanced communication technology and infrastructure (4G, 5G, LoRA, ZigBee, etc.)
(4) Big Data storage and solutions
(5) Analytics/data science platforms
(6) Advances in sensor and actuator technologies
(7) Energy saving equipment
(8) Predictive maintenance
(9) Smart cities
(10) Interconnected devices/assets and people

The disruptions are further accelerated with the expounding developments in the area of machine learning (ML) and artificial intelligence (AI). Enterprises are increasingly focused on leveraging data-driven planning via AI/ML applications. To ensure the success of IIoT and AI-enabled manufacturing systems, there needs to be focus on the drivers, namely, people, process, and technology. A conceptual schematic for success drivers is presented in Figure 1. Key applications in the manufacturing space include (but are not limited to) the following:

- Adaptive control systems for tracking and control of manufacturing process.
- Robots on the shop floor/assembly line replacing the manual tasks.
- Automation systems with end-to-end (i.e., raw material to final product) value add.
- Real-time scheduling and optimization systems.
- Cyber-physical security systems.

- Early warning health–safety–environment (HSE) systems for workers' and asset safety.

The IIoT comprises interconnected sensors, instruments, and other devices networked together with computers' industrial applications, including manufacturing and energy management (Figure 2).

Over 72% of the organizations consider IIoT to significantly contribute to reduction of unplanned downtimes (Zheng *et al.*, 2020). As per recent surveys, around 3–8% (Lin, 2019; Tim, 2020; Balo, 2016) of the

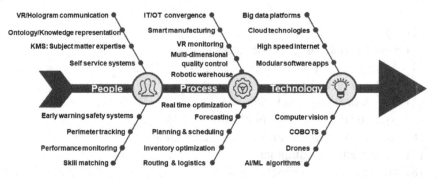

Figure 1.   Drivers for ensuring success with IIoT & AI-enabled manufacturing systems.

IIoT & AI enabled Manufacturing Systems: Conceptual View

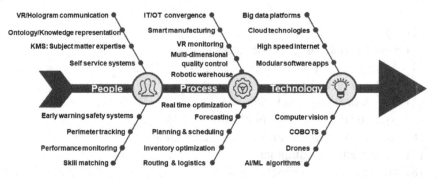

Figure 2.   Conceptual view of IIoT and AI-enabled manufacturing systems.

world's manufacturing systems are connected to network. The success of IIoT and its adoption in the German industry are surveyed by Arnold & Voigt (2019). These surveys indicate the initial success of the IIoT & AI-enabled manufacturing systems and raise hopes for the immense potential offered by them.

The Industrial Internet of Things (IIoT) was conceptualized as a value driver from the industry and then had backward integration with academics. There are over 3000 journal papers available dated till 2018 on smart factories and Industrial Internet of Things (Madakam & Uchiya, 2019). The authors of this chapter survey some of the literature and importantly group them into the different pillars contributing to a digital/smart manufacturing road map.

Ghosh *et al.* (2020) give a crisp overview of the theoretical concepts pertaining to IIoT. The authors' work complements theirs by presenting an underlying sensory system/building blocks and the use case applications. Yang *et al.* (2016) conceptualize the IoT technologies as a seamless connection between the physical systems and cyberspace. Authors conceptualize real-time control with AI systems which power the technology with decision-making algorithms and real-time control/actions.

2. Conceptual Framework: IIoT as Sensory Organs of Manufacturing Systems

In this chapter, the authors propose the notion of sensor technologies and advanced decisioning capabilities in the IIoT systems as analogous to the

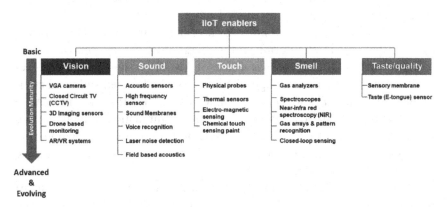

Figure 3.   Enablers of IIoT sensory system.

"sensory organs". The success of IIoT in manufacturing systems is based on a strong foundation of the data acquisition, collection, and processing systems. Figure 3 shows the evolution of IIoT sensory systems.

2.1. *Eye: Sense of vision/sight*

Sense of sight is a key capability for the manufacturing systems as that they can "view" objects, and/or resources and hence track, plan, and react/pro-act accordingly. The sight capabilities are provided by IIoT industrial sensors ranging from a simple VGA camera to drone-based camera(s) with advanced image and real-time videos and advanced AR/VR (augmented reality/virtual reality) systems.

Johnson *et al.* (2019) debate the visual inspection skills between humans and IIoT. They propose the capture of "tacit knowledge" as a missing link between digital technology and human cognitive abilities. Hence, capturing and learning from the "tacit knowledge" to improve the sight/visual capabilities of the manufacturing systems are important. The authors discuss a case study of visual inspection with tacit knowledge used in the context of a UK-based aerospace components manufacturer. The tacit knowledge complements digital technology to provide interpretation capabilities, especially for non-standard dimension system.

Mar *et al.* (2011) present a visual system for inspection of solder joints in an industrial context. They use a "digital" camera as an IIoT device to enable remote vision capabilities for the manufacturing system.

For a detailed survey of visual technology in the manufacturing systems (specifically in the semiconductor industry) context, interested readers are referred to a survey paper by Huang & Pan (2015).

Alhayani & Llhan (2021) discuss a visual module for manufacturing systems that aids in monitoring and is energy efficient in terms of design.

Siedler *et al.* (2021) discuss a case of AR/VR systems used for a factory layout and machine changes. The senses of advanced sight and comprehension with AR/VR will be a game changer for time-sensitive agile manufacturing scenarios. Using an effective combination of methods and tools, the authors show that the implementation time of engineering changes can be shortened and the likelihood of cost-intense planning errors can be eliminated/reduced.

Kim *et al.* (2018) propose the use of AR/VR systems in the smart manufacturing context. This enables factory workers, planners, and maintenance personnel to plan and use the resources effectively while sticking to the protocols. Any breaches in the protocol can be alerted in near-real time, hence minimizing the possibility of accidents and unwanted events.

Dhanalakshmi *et al.* (2021) discuss the growth in Industry 4.0 with the advent of AR/VR systems. The manufacturing sector's profit grew non-linearly owing to Industry 4.0 data collection systems have been revamped with AR/VR-based capabilities.

2.2. *Ear: Sense of sound/hearing*

Acoustic data are high-frequency unstructured data, which when combined with state-of-the-art machine learning/artificial intelligence techniques will provide interesting and valuable insights for manufacturing assets in near-real time.

Acoustic data form a valuable source of industrial data. The sound (data) originating from various assets (equipment/personnel) when captured and processed through IIoT smart technologies can be of immense value. For example, the data can be used for anomaly detection of industrial compressors and turbines as indicated in the paper by Serizawa & Shomura (2019). The authors developed an audio streaming data-sensing IIoT system and illustrated a machine classification and anomaly detection use case.

Shevchik *et al.* (2019) employ acoustic sensor data in the IIoT context coupled with machine learning algorithms for process control of manufacturing equipment.

Talmoudi *et al.* (2019) show that IIoT sound sensors usage can lower the cost of predictive maintenance. They use data transmission between the sensor nodes and the server from a DC motor with failure mode(s) induced by the variation of speed and other parameters. Their failure prediction is accurate enough for industrial-scale adoption. The early warning signals on deviation in the operations mode could be tracked over a day window of the actual occurrence of the event.

Fisher *et al.* (2017) demonstrate the use of laser-based acoustic sensors for process monitoring and control. Laser-based sensing can help in the industry environment which could be contaminated with debris and dust. Traditional sensors would not be suited in this contaminated environment.

Lee *et al.* (2006) use acoustic emissions for a precision manufacturing process in ultraprecision diamond turning, where the accuracy rates ought to be over and above 99%.

Kim *et al.* (2021) employ sound sensors for remote data acquisition from disparate manufacturing resources. They employ a convolutional neural network (deep-learning) algorithm for processing the data and classify the machine state to anomaly vs normal operations zone. Their model achieves accuracy to the north of 90%.

2.3. *Skin: Sense of proximity/touch*

The sense of proximity/touch forms an important data element in the smart manufacturing context to detect/alert any breach of operating geographic boundaries (referred as "geo-fencing"). The scale of touch can range from a few meters to a precision of micro/nanometers, thus catering to a widespread range of practical industrial systems. The touch sensors can range from simple probes to electro-magnetic fields. It is worth noting that ever since the beginning of the COVID-19 pandemic, there has been increasing industrial R&D (research and development) focus on non-touch proximity sensors, leveraging the electro-magnetic fields, radiation, etc.

Rahman *et al.* (2016) emphasize the importance of touch sensors in the Human–Machine Interfaces of IIoT systems. They present an additive manufacturing approach for touch sensors capable of sense precision of 50 μm. Ntagios *et al.* (2020) show a proof-of-concept of a futuristic robot with touch sensors embedded in a robotic hand that is ready to be deployed in an IIoT-based manufacturing system.

2.4. *Nose: Sense of smell/odor*

Fragrance/odor detection is important in hazardous manufacturing environments. Hazardous gases are layered with fragrance/odor chemical agents to enable their detection via industrial sensors and raise an alert on their detection to the resources within the impact perimeter.

Kavitha & Srinivasan (2020) present an eNose, i.e., IoT-enabled electronic nose, to mimic the function of a human nose in detecting smell. The eNose comprises an array of gas sensors embedded with a pattern reorganization module. The authors propose examples of eNose usage in an industrial manufacturing setting. This is an interesting pilot-stage exploration.

Wang *et al.* (2018) conceptualize a futuristic AI system with AR/VR capabilities that brings in "smell" with augmented reality in smart manufacturing systems. The authors hypothesize that a smell sensor with AR/VR helps complement operator(s) where the work environment has limited flexibility (possibly hazardous operations).

2.5. *Tongue: Sense of quality/taste*

Quality is the differentiating factor for manufacturing systems. IIoT systems are equipped with ways to identify substandard quality products and processes and initiate a corrective/preventive action as deemed appropriate and cost effective. Measuring taste/quality is more subjective and harder to quantify.

An electronic tongue is a recent/evolving area of research in the medical industry. The concept of taste of substance to ensure safety, quality, and adherence to the preferences of taste is a game changer for food manufacturing and pharma industries. Guedes *et al.* (2021) show the use of an electronic tongue as a sensor to evaluate "taste", applicable in the context of the pharma/medical industry. Jayaram (2016) conceptualizes a six-sigma lean approach for global supply chains. IIoT-based quality processes and sensing tools/methods help craft an optimal process flow to reduce and/or eliminate defects and wastage from the supply chain.

3. AI as the Mind and Central Nervous System in Manufacturing Systems

AI forms the central nervous system and the brain of the IIoT-enabled manufacturing systems. While the IIoT as illustrated in this section performs the role of sensory organs, AI-based algorithms form the brain of the manufacturing systems.

There is a vast amount of literature on AI-enabled manufacturing systems. Interested readers are referred to a 1990 paper by Parunak & Dyke (1990). There is a recent comprehensive review paper on AI applications in manufacturing by Issac Kofi Nti *et al.* (2021). Prasad and Choudhary in a recent paper (Prasad & Choudhary, 2021) presented a review on the state of the art of AI, the evolution of the AI landscape, and its future state.

In the context of the current chapter, the authors review the relevant papers which cover the IIoT.

Shah *et al.* (2020) present a detailed view of feature engineering and the underlying AI algorithms and their differentiators in a broad industrial application point of view. They emphasize the role of IIoT sensors in the manufacturing process and control, while AI algorithms help in crafting an optimal course of action(s).

3.1. *Unsupervised learning*

Unsupervised learning is a fundamental AI/ML technique (Figure 4), where the model is intended to discover hidden relations/patterns from the data. There is no explicit target/output identification before the model is fitted, i.e., there is no training data to identify the input vs output. In other words, the data used for modeling are "unlabeled", and the outcome of the unsupervised learning model would be a "label" for the data.

In a recent paper, Cheng *et al.* (2021) discuss an artificial bee colony unsupervised learning algorithm for identification of non-value-added activities in the manufacturing context. The authors present their algorithm for the forging industry with heterogeneous parallel machines.

Wocker *et al.* (2020) use unsupervised learning algorithms to analyze type mixes and related process performances in a flexible manufacturing system (FMS). The authors came up with an optimal master production schedule that considers the timing of maintenance activities, henceforth increasing the overall system availability.

Chen *et al.* (2020) employ an unsupervised learning algorithm for analyzing sensor data from a semiconductor manufacturing industry

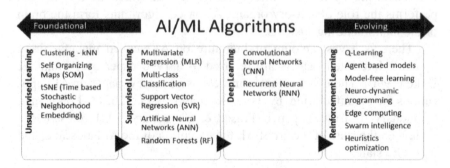

Figure 4.   Landscape of AI/ML algorithms.

to identify anomalies. They use an advanced algorithm, namely, auto-encoders, on the sensory data for the anomaly detection.

Kolokas *et al.* (2020) propose a prognostic fault detection algorithm for manufacturing systems employing unsupervised learning models. Dogan & Briant (2020) elaborate on the list of machine learning and data mining algorithms for manufacturing systems. They detail clustering methods which are a subset of unsupervised learning. Ren *et al.* (2020a) present a k-Nearest Neighborhood (kNN) clustering model for a biopharma manufacturing case. They show the application of a kNN model for process diagnosis and performance monitoring for Biogen (pharma manufacturer).

3.2. *Supervised learning*

Supervised learning forms the core of AI/ML algorithms. In a supervised learning framework, the data used for model building are split into input vs output. Using a subset of the data with the input–output mapping, the supervised learning algorithm model is employed to arrive at a "best fit" model that maps the inputs to an output.

Depending on the variable type of the output, the supervised learning models are further categorized into two: (a) Regression models — the output variable is a real number; (b) Classification models — the output variable is a whole number. Most of the classification problems tend to have binary outcomes and hence are commonly referred to as "binary classification". There are a vast variety of applications of the supervised learning algorithms in the context of manufacturing systems. Yu *et al.* (2020) apply supervised learning algorithms for classification of manufacturing data to a fault/no-fault state, and hence assess the important factors that impact the outcome. Izagirre *et al.* (2021) use vision-based IoT data in combination with the supervised learning methods to predict robotic accuracy degrade and the factors affecting this.

3.3. *Deep learning*

Deep learning is an emerging area in the AI/ML space. Deep learning deals with vast amounts of unstructured data, e.g., images, videos, speech/audio, and text, and imitates the workings of the human brain in processing data and creating patterns for use in decision-making via an adaptive and complex artificial neural network-based model.

Bhuvaneswari *et al.* (2021) give a detailed review of the evolution of deep learning from conventional machine learning, with a focus on manufacturing systems applications.

Kotsiopoulos *et al.* (2021) introduce deep learning in the context of Industry 4.0. They review important algorithms in deep learning at a high level, complementing the case of smart grids. Meng *et al.* in their 2020 paper (Meng *et al.*, 2020) present a high-level view of the machine learning algorithms in the context of additive manufacturing systems. As a connected research, Li *et al.* (2020) use a deep-learning approach for analyzing quality in additive manufacturing systems. The authors demonstrate a high-accuracy output identifying quality issues based on low-resolution image data obtained from a metal additive manufacturing system. Their approach demonstrates considerable cost and time savings while achieving superior quality.

3.4. *Reinforcement learning*

Reinforcement learning (RL) is a hybrid AI algorithm blending in the best of supervised and unsupervised learning algorithms and using it to create actionable insights. Thus, RL can be viewed as an adaptive, intelligent algorithm that enables AI actions in near-real time. Xia *et al.* (2021) propose a novel application of deep reinforcement learning, enabling intelligence in the digital twin manufacturing systems. This enables the adaptive learning of agents and improves decision-making abilities via AI algorithms for manufacturing systems.

Oliff *et al.* (2020) present a reinforcement algorithm applied for a human–robot interface communication/action in a manufacturing environment. They do an interesting experiment analyzing robot performance using reinforcement learning against the performance variation of a human operator. These kinds of pilot studies can be viewed as baby steps for the adaption of RL at scale for world-class manufacturing systems. Another interesting case study of application of RL in manufacturing systems is by Paraschos *et al.* (2020). They demonstrate the use of Reinforcement Learning (RL) for manufacturing systems in production quality and maintenance planning. Here, the computational agents for RL are collaborating/competing against two objectives and try to achieve the best possible outcome.

Self-repair/self-healing manufacturing systems are futuristic research areas with the RL. One such recent example is presented in a paper by

Epureanu *et al.* (2020). They examine market and product quality and use RL to create levelized self-repair strategies. RL is deployed in strategy selection, thus optimizing the status of the system and increasing the performance.

The authors believe that self-healing systems are a good start, which are currently restricted to the laboratory/pilot stage. The wide-scale manufacturing industry will adapt these in a span of 5–10 years.

4. Manufacturing IIoT Landscape: Use Case Perspective

IIoT is an amalgamation of industrial automation, control systems, and IoT systems (Bansal *et al.*, 2021). In this paper, the authors categorize the manufacturing systems into three connected areas, namely, product design, logistics, and core manufacturing processes (Figure 5).

4.1. *Product design*

New product development and innovation are key differentiators for a firm and contribute to its competitive advantage (Klintong *et al.*, 2012).

Use cases in Manufacturing Systems

Product Development	**Warehousing & Logistics**	**Core Manufacturing**
New product development	Quality management	Asset management & optimization
Rapid prototyping & testing	Sourcing selection	Predictive maintenance
Customer/market sensing	Fleet management	Optimal production
Feature/product importance	Real-time tracking & alerting	Process & product quality monitoring, control
Preference elicitation.	Network design	Forecasting and control
Product demand diffusion	Drone-based delivery	Operations management
	AS/RS systems	
	Robotic warehouses	

Figure 5.   Use cases in IIoT-enabled manufacturing.

Nozaki *et al.* (2017) provide an application of AI in product design. They discuss a case of PCB (printed circuit board) design. They show how AI can be applied to elicit human responses by leveraging human experience and know-how and thereby designing better products.

Li *et al.* (2017) give a survey of AI applications in intelligent manufacturing. They mention industrial applications in China and state how AI applications have proven to add value in a real-world manufacturing scenario.

Seetharaman & Sharma (2019) present a view of customer expectations in the context of IIoT relevant to the product design. The authors examine fundamental capabilities: connectivity, big data, advanced analytics, and application development. They examine IIoT potential synergies between the 4Ms of manufacturing, namely, man, machine, material, and method.

Rao *et al.* (1999) gives a detailed survey of the application of AI and Expert Systems (ES) in new product development. He categorizes the applications into five areas, namely, expert decision support systems for NPD project evaluation, knowledge-based systems (KBS) for product and process design, KBS for (quality function deployment) QFD, AI support for conceptual design, and AI support for group decision-making in concurrent engineering.

Choy *et al.* (2004) provide a case study of Honeywell Consumer Product Limited (at Hong Kong), wherein they discuss an intelligent supplier relationship management system comprising hybrid case-based reasoning combined with Artificial Neural Networks (ANNs) for benchmarking potential suppliers. As a result of the application of AI techniques, the authors demonstrate reduction of the outsourcing cycle time in searching for the potential suppliers and allocation of the orders.

Soltani & Pooya (2018) use AI for the prediction of a new product in the context of the food industry. They show a relation between the actions of CXOs in developing the new products, while surveying over 250 companies.

Lopez & Casillas (2013) survey use cases of intelligent systems in the context of industrial marketing. They indicate that the major applications of AI in the context of industrial marketing are focused on (a) management and (b) pricing decisions. Further, the authors mention the increase of AI systems in industrial marketing-related decisions.

Kwong *et al.* (2016) discuss an AI-based methodology for integrating affective design, engineering, and marketing for defining the design specifications of new products. Their proposed methodology is useful in the early stages of design and highlights concerns in the three processes, namely, affective design, engineering, and marketing, simultaneously. They present a case study of an electric iron design for demonstrating the effectiveness of their proposed methodology.

Feyzioglu & Ozkan (2007) emphasize the uncertainty and riskiness involved in new product development. In addition, they highlight the pressure of the launch of products, warranting a robust evaluation approach for new product development projects. They use neural networks and fuzzy logic to evaluate new product development projects and propose go/no-go decisions. They give a systematic AI approach improve the quality of the decision-making and to increase the probability of success in New Product Development (NPD) under uncertainty by introducing an iterative methodology.

4.2. *Warehousing and logistics*

Warehousing and logistics play key roles in the manufacturing systems. Warehouses serve as the stocking points for the raw material, intermediate goods, and finished products. They are key to managing seamless operations in core manufacturing, i.e., ensuring the availability of right product inputs to the production system and right product outputs to the customers/delivery points. Warehouses are important for normal operations, and they come to the rescue in times of physical disruptions by way of utilizing the inventory in the warehouse to act as a cushion until the disruption effects are mitigated. Logistics complements the warehouse and manufacturing operations. It transports the inputs from vendors to manufacturing facilities via inbound logistics. It transports the outputs from manufacturing facilities to intermediate and/or end customers through outbound logistics. The authors refer to the below schematic view proposed by the Chartered Institute of Procurement and Supply (CIPS) illustrating the links between warehouse, logistics, and manufacturing (Figure 6).

Warehousing and logistics connect core manufacturing with external stakeholders, e.g., supply side (vendors), product designers, research and

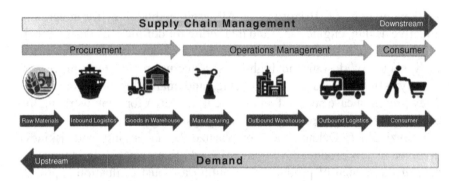

Figure 6.   Integrated manufacturing systems (core manufacturing, warehouses, and logistics).

Source: CIPS (2020).

development, demand side (customers), warehouses, and retailers. They involve multiple physical touch points/interaction nodes with warehouse assets and logistics assets. Often, these touch points are in the thousands, thus integrating the manufacturing systems into warehouse and logistics, as the last mile or penultimate mile is a gigantic operational and technical problem. In this context, the authors cite Amazon, which grew into a multi-billion-dollar enterprise through a strong foundation and continued innovation in warehousing and logistics. Amazon Robotics is a key differentiator in achieving Amazon's supply chain delivery excellence, managing millions of products with variety in scale and scope (AMA). Mobile technology integrated with the warehousing and logistics is a breakthrough innovation enabling the tracking of objects in motion, while ensuring connectivity to assets in the network.

Barreto *et al.* (2017) give a comprehensive overview of the Industry 4.0 implications for the logistics industry. They emphasize the technological and functional challenges posed by the IIoT and how the logistics industry should adapt to these by ensuring the right product at the right place with the right speed.

Zhang *et al.* (2018) show smart intra-factory logistics powered by IoT to connect production and logistics in a job-shop environment. They demonstrate IoT-based logistics for a Chinese engine manufacturer. The results of their study are encouraging for IIoT logistics, and they show reduced overall manufacturing time and energy consumption.

Stefano *et al.* (2020) detail the technology side of IIoT for logistics. They present an EtherCAT network-based case of a gantry robot in an automated warehouse. They show optimized efficiency via IIoT in logistics.

Lee (2018) implement an IIoT-enabled robotic warehouse system. Their system comprises autonomous mobile robots deployed in a warehouse for picking and moving goods. Their study indicates that IIoT-enabled smart warehouses which maximize the floor space and work force utilization simultaneously improve the logistics efficiency.

Zelbst *et al.* (2019) discuss an emerging concept of blockchain coupled with an RFID-based IIoT system and demonstrate how the transparency of the logistics value chain improves. RFID forms an early yet passive and a promising sensing technology in the logistics space.

Drone-based technology to track the movement and/or deliver goods is an emerging area in the industrial manufacturing systems. For a detailed overview of opportunities with drone-based systems for logistics, readers are referred to the paper by Maghazei & Netland (2019) and an interesting article by Floreano & Wood (2015).

Olivares *et al.* (2015) present a drone-based application for internal factory logistics in a manufacturing assembly context. The drone movement is optimized via the traveling salesman problem and solved via a genetic algorithm. Thus, this becomes a case for IIoT with AI for drone-based logistics.

4.3. *Core manufacturing*

Gujara *et al.* (2020) show a use case of IIoT in the manufacturing system. They show the advantages of IIoT in reducing the overall production cost and in improved utilization, further achieving an overall optimization of manufacturing and planning.

4.3.1. *Planning and scheduling in manufacturing systems*

AI applications in planning and scheduling date back over six decades. Li & Zhang (2020) review the application of virtual simulation technology and simulation-based optimization for pre-planning in a production line layout, plant resource allocation and scheduling, production process, warehouse, and logistics. The authors emphasize the (a) growing importance of reinforcement learning and AI applications in manufacturing planning and scheduling, and (b) human modeling aspects in scheduling. They mention scheduling in cloud systems and digital twin as emerging areas of AI applications.

Arinez *et al.* (2020) survey AI applications in manufacturing. They present (i) a review of the state-of-the-art applications of AI to representative manufacturing problems, (ii) a systematic view for analyzing data and process dependencies at multiple levels that AI must comprehend and conclude with, and (iii) identify challenges and opportunities for leveraging and facilitating future development of AI to better meet the needs of manufacturing.

Krause (2020) uses an AI-based discrete event simulation model for manufacturing schedule optimization. The simulation model is based on an actor-critic algorithm (commonly used in the context of reinforcement learning). The author suggests the usage of AI to connect the gaps of data quality in the information systems connected to the Enterprise Resource Planning (ERP) system, thus making the AI-based simulation more practical from an industrial application point of view. The author demonstrates the working of the simulation model for over 300 products for a planning horizon of 24 months.

Fahle *et al.* (2020) present a recent review of AI applications in the manufacturing field operations context, based on the ML model(s) used for underlying applications. They identify two gaps, namely, assistance systems and learning factory training.

4.3.2. *Inventory management systems*

One of the very first applications of Internet of Things in inventory management started with passive sensors as an extension of the Radio Frequency Identification Device (RFID) tags.

Grangier (Staff) in its report discusses applications of IoT in inventory management as follows:

- Real-time data from IoT devices help create a more streamlined supply chain. IoT helps in capturing and collating real-time inventory movement data, tracking the positions of a shipment, and thereby creating a transparent supply chain.

- IoT devices provide precise location monitoring and enhanced inventory tracking. IoT provides location monitoring, further integrating it with the inventory accounting systems.
- RFID and touchless data collection from IoT devices help create error-free processes and reduce shrinkage. With the emergence of COVID-19, the emphasis is on touchless data collection, verification, and eventual consumption (e.g., report or analytics services). IoT enables this effectively and efficiently, keeping it safe and healthy.
- IoT devices help improve logistics and warehouse efficiency.
- Predictive analytics from IoT devices helps identify potential risks and enhances performance measurement.
- IoT devices enhance fleet inventory management.

Jayaram (2017) use an integrated IIoT system for (a) inventory management, (b) demand forecasting, and (c) enterprise automation

- In his thesis, Futardo (2020) discusses a minimum viable product (MVP) requirement for IIoT-based inventory management. This concept of MVP plays a key role in scoping a big implementation to achieve practical results in a faster and efficient manner.
- Bottani *et al.* (2019) use an artificial intelligence-based framework to support decision-making in wholesale distribution. They minimize wholesaler out-of-stocks by jointly formulating price policies and forecasting retailer's demand.
- Oh (2019) uses an AI model for a transactional relationship in supply chains that enables efficient inventory management and timely product supply, contributing to maximizing corporate profits.
- Baryannis *et al.* (2019) propose AI models for supply chain risk model applications. The authors emphasize the complexity and characteristics of supply chain risk that makes it a potential application area for AI models.

4.3.3. *Predictive maintenance*

Predictive maintenance has been one of the largest applications of the IIoT. It is estimated that the IIoT usage would cut down maintenance cost. Applying IIoT and AI techniques has reduced the maintenance speed by over 15% McKinsey, 2021 survey). When viewed from a predictive maintenance angle, the IIoT landscape comprises five layers:

- Descriptive (past).
- Diagnostics (understanding of present through past data).
- Predictive (forecasting).
- Prescriptive (optimization).
- Autonomous (self-healing actions).

In a recent paper, Zheng *et al.* (2020) estimate that the losses due to unplanned downtime for 1 year across an asset-intensive manufacturing industry would be in the order of 50 billion USD. This loss can be cut down significantly via the connect between IIoT and AI. The authors demonstrate efficiency of the IIoT systems, particularly in the predictive maintenance of industrial equipment.

Ren *et al.* (2018) present an IIoT application coupled with deep learning to predict remaining useful life of bearings. They show IIoT to improve understanding of the usage patterns of bearing, their failure characteristics, etc. As a continuation to this research, the authors extend this further and provide a cloud-based application for predicting the remaining useful life (Ren *et al.*, 2020b).

5. Futuristic/Emerging Applications: IT–OT Integration, Edge Computing, and Technological Aspects

In the context of intelligent manufacturing, Digital Twins emerged as a key disruptor. This has enabled the connect from the (shop) floor to the board, enabling faster and more efficient decisions. The key to enablement of real-time decisioning for manufacturing systems is via edge computing.

Alexakos *et al.* (2018) examine the integration aspects of the shop floor and plant layer systems. They further discuss the hierarchical integration among three layers of enterprise, namely, ERP (Enterprise Resource Planning), Manufacturing Execution Systems (MES), and the Shop floor.

Zhou *et al.* (2020) propose three key enabling technologies, namely, digital twin model, dynamic knowledge bases, and knowledge-based intelligent skills for supporting above strategy are analyzed. They equip the system with the capacities of self-thinking, self-decision-making, self-extraction and self-improving in real world manufacturing context.

Aditya & Srikanth (2017) discuss an application of Digital Twin in vehicle manufacturing. They conceptualize that Digital Twin creation provides untapped opportunities in terms of layout design, space utilization, and equipment efficient operations.

The initial results on leveraging edge computing for industrial manufacturing systems are encouraging. This is still an evolving area both in terms of technology and adoption in business. To the best of the authors; knowledge, there is no scalable breakthrough yet for deployment in a large-scale manufacturing industrial system. Zaho *et al.* (2020) discuss a pilot on edge computing in industrial manufacturing networks.

5.1. *Technology aspects*

Technology plays a key role as enabler for the smart technologies that could be leveraged in manufacturing systems. The major focus of this chapter is on use case aspects; hence, the authors intend to cover a few relevant works on technology aspects in the IIoT.

Madakam & Uchiya (2019) provide a detailed view of the IIoT principles, processes, and protocols, which complement with the real-world IIoT implementation case studies in the Japanese industry, e.g., Hitachi, Toyota, and Mitsui. A key challenge in the technology for manufacturing systems is interoperability. For instance, Lia *et al.* (2017) emphasize the need for interoperable technology. The protocols common in the IIoT space are as follows: IIoT, IoE, HTTP, REST, JSON, MQTT, OPC UA, and DDS. There are two main protocols that gained popularity in the IIoT from the end user viewpoint, namely, Message Queuing Telemetry Transport (MQTT) and Constrained Application Protocol (CoAP). Lu *et al.* (2020) discuss the IIoT standards commonly used in smart manufacturing. Traditionally, industrial systems are connected via SCADA (Supervisory Control and Data Acquisition). SCADA systems use proprietary protocols resulting in interoperability challenges. MODBUS TCP is an open de facto standard and is used for some automation and telecontrol systems. MQTT is chosen as the event-based, publish–subscribe protocol for most of the IIoT protocols (Jaloudi, 2019). Until early 2015, the manufacturing industry did not perceive a clear value out of IIoT. Their main apprehension was that the manufacturing industry was already using the sensors and actuator devices, and how the value would impact when connected to the internet was not evident (Gilchrist, 2016).

Boyes *et al.* (2018) attempt to define the Industrial Internet of Things (IIoT) from a technical point of view and analyze several taxonomies of IIoT. Their work emphasizes the architecture aspects and security/vulnerabilities in the IIoT space. The authors categorize IIoT literature into two strands: a) kinds of technologies used in an IIoT setting and (b) the use cases for IIoT.

There are six different IIoT taxonomies as summarized as follows:

- The device-centric taxonomy — focuses on devices and interfaces.
- The IoT stack-centric taxonomy — focuses on service data and data exchange between decentralized and non-standard systems.
- The IoT sensor taxonomy — focuses on the underlying sensor/actuator technology.
- The IoT-based smart environment taxonomy — focuses on environmental and system perspective, i.e., the big picture of IIoT.
- The IoT architecture taxonomy — this is an evolving approach blending technical and functional architecture. Hence, it is suitable for the modern techno-functional business.
- The Industrial Internet of Things taxonomy — this is more recent, focusing on emerging aspects like cyber-physical security, carbon footprint, resilience, and reliability of industrial systems. Here, the focus is on practicality and scalability.
- The domain or sector-based IoT taxonomies — focus on the underlying vertical/domain, e.g., automotive, defense, chemicals, energy, semiconductors, and heavy manufacturing.

Ungurean *et al.* (2016) propose an IoT architecture designed around the Data Distribution Service for Real-Time System (DDS) middleware protocol. The architecture is scalable at the middleware level; the DDS can be replaced by other middleware systems thus covering the interoperability aspects.

Kan *et al.* (2018) propose a stochastic algorithm for large-scale IIoT fault diagnosis. They use a combination of a dynamic warping algorithm and a stochastic network embedding algorithm for machine state classification and operations signatures.

Loske *et al.* (2019) focus on an important topic of authentication in IIoT. This is an emerging research area for industry. Traditional authentication methods of IIoT are constrained by power and resource bandwidth. They suggest context-aware authentication for machine-to-machine authentication and propose high-level use cases. Interested researchers can pursue this as a future area of research.

Lorenzo *et al.* (2020) present a detailed survey of IIoT protocols while placing emphasis on the security risk in IIoT integration. They survey a total of 33 protocols, standards, and buses used in an IIoT environment and propose a Vulnerability Analysis Framework that enables qualification and quantification of over 1000 IIoT vulnerabilities.

6. Conclusions and Directions for Future Research

Usage of IIoT is prone to long-term impacts in the physical world. IIoT will disrupt several ecosystems. A key development in this aspect is that economies are redesigning their respective country's manufacturing roadmap and its contribution to the overall nation's GDP via IIoT-enabled manufacturing. IoT is estimated to contribute (directly/indirectly) to two-thirds of the global GDP (Gilchrist, 2016). Some recent examples from Asia are Make in India, Make in China 2025, Smart cities, and the Japanese Industrial value forum (Madakam & Uchiya, 2019). Arnold & Voigt (2017) conducted a survey among German manufacturers to understand IIoT adoption and its perceived impact on business ecosystems. They use the following dimensions to characterize an ecosystem: customers, suppliers, external organizations, and research institutions. They conclude that IIoT adoption is associated with greater openness of manufacturers toward participants of all analyzed ecosystem dimensions.

A major impediment to the adoption of IIoT happens to be cybersecurity issues. For successful usage of IIoT, industries must redesign their cyber-physical systems. Businesses require holistic solutions that are transparent and provide the ability to trace data and pinpoint failures at both technical and organizational levels. As the complexity in IT infrastructures increases with billions of IIoT devices, encompassing a host of endpoints, cybersecurity detection and remediation with speed dictate the IIoT's full success (Herbert, 2019). Hassan *et al.* (2020) use an adaptive trust boundary protection framework along with a deep-learning-based semi-supervised learning model to identify any network invasion/attack of IIoT systems. Their framework is robust and currently piloted to suit complex world-class manufacturing system requirements. Zhang *et al.* (2021) raise an important yet thoughtful question on the ethics and governance of AI systems. Abuhasel & Khan (2020) propose a secure IIoT framework for smart resource management in manufacturing systems. With the advent of cyberattacks on manufacturing systems, energy distribution networks, and other capital-intensive asset industries, future research should focus on strengthening the IIoT network with secure protocols and systems that can track, monitor, and proactively neutralize the threats to manufacturing systems.

6.1. *Directions for future research*

IIoT and AI-enabled manufacturing systems are evolving at an exponential pace, simultaneously generating multidimensional research and industrial application opportunities. The authors of this chapter highlight for interested readers some of the areas that can be considered for future research.

- The aspects of industrial cyber-physical security, real-time analytics, and edge computing are evolving areas of research that are poised to take the manufacturing systems to their next level of growth and value. The authors suggest that the unintended consequences of AI and technology could pose a larger threat to the world, and a systematic means of identification and elimination of these threat factors should be considered at the design stage of the AI systems.
- A multi-billion-dollar industrial research problem is on-boarding considerations for older machinery/assets to IIoT. This would investigate and propose some of the novel possible ways of retrofitting IIoT sensors and connectivity to the internet in an economical and least invasive manner.
- Interoperability design for IIoT devices, establishing and standardizing communication protocols across heterogeneous vendor manufacturing of IIoT devices while triaging the cyber-physical security aspects with manufacturing economics, is a major area for industrial research.
- In terms of technical design considerations, an open area of research is designing IIoT sensors with low power consumption, while still

adapting to the latest and hottest communication and internet technologies, e.g., 5G readiness for IIoT.
* A logical extension of the authors' work is to create a theoretical taxonomy framework and an ontology-based representation of the IIoT and AI-enabled manufacturing systems.

Acknowledgments

The authors thank the two anonymous reviewers and the editor for their valuable inputs in improving the overall organization of the chapter. The second author expresses his thanks to Dr. Abdulrahman Alattas (ADNOC) and Mohamed Saleh Al Katheeri (ADNOC) for their constant support and encouragement.

References

10 ways AI is improving manufacturing in 2020. (2020). https://www.forbes.com/sites/louiscolumbus/2020/05/18/10-ways-ai-is-improving.

Abuhasel, K. A. & Khan, M. A. (2020). A secure industrial internet of things (IIoT) framework for resource management in smart manufacturing. *IEEE Access*, 8, 117354–117364.

Aditya, V. Y. & Srikanth, P. (2017). IIoT-enabled production system for composite intensive vehicle manufacturing. *SAE International Journal of Engines*, 10(2), 209–214.

Alexakos, C. Anagnostopoulos, C. Fournaris, A. Koulamas, C. & Kalogeras, A. (2018). Iot integration for adaptive manufacturing. In *2018 IEEE 21st International Symposium on Real-Time Distributed Computing (ISORC)* (pp. 146–151). IEEE.

Alhayani, B. S. A. & Llhan, H. (2021). Visual sensor intelligent module-based image transmission in industrial manufacturing for monitoring and manipulation problems. *Journal of Intelligent Manufacturing*, 32, 597–610.

Arinez, J. F. Chang, Q. Gao, R. X. Xu, C. & Zhang, J. (2020). Artificial intelligence in advanced manufacturing: Current status and future outlook. *Journal of Manufacturing Science and Engineering*, 142(11), 8.

Arnold, C. & Voigt, K. (2017). Ecosystem effects of the industrial Internet of Things in manufacturing companies. *Acta Infologica*, 2(1), 99–108.

Arnold, C. & Voigt, K. (2019). Determinants of industrial internet of things adoption in German manufacturing companies. *International Journal of Innovation and Technology Management*, 16(06), 1950038.

Balo, F. (2016). Internet of Things: A survey. *International Journal of Applied Mathematics, Electronics and Computers*, 12, 104–110.

Bansal, M., Goyal, A., & Choudhary, A. (2021). Industrial Internet of Things (IIoT): A vivid perspective. In *Inventive Systems and Control* (pp. 939–949). Springer.

Barreto, L., Amaral, A., & Pereira, T. (2017). Industry 4.0 implications in logistics: An overview. *Procedia Manufacturing*, 13, 1245–1252.

Baryannis, G. Validi, S. Dani, S. & Antoniou, G. (2019). Supply chain risk management and artificial intelligence: State of the art and future research directions. *International Journal of Production Research*, 57(7), 2179–2202.

Bhuvaneswari, V., Priyadharshini, M., Deepa, C., Balaji, D., Rajeshkumar, L., & Ramesh, M. (2021). Deep learning for material synthesis and manufacturing systems: A review. *Materials Today: Proceedings*, 46(1). DOI: 10.1016/j.matpr.2020.11.351.

Bottani, E., Centobelli, P., Gallo, M., Kaviani, M. A., Jain, V., & Murino, T. (2019). Modelling wholesale distribution operations: An artificial intelligence framework. *Industrial Management & Data Systems*, 119(4), 698–718.

Boyes, H., Hallaq, B., Cunningham, J., & Tim Watson. (2018). The industrial internet of things (IIoT): An analysis framework. *Computers in Industry*, 101, 1–12.

Chen, C., Chang, S., & Liao, D. (2020). Equipment anomaly detection for semiconductor manufacturing by exploiting unsupervised learning from sensory data. *Sensors*, 20(19), 5650.

Cheng, C., Pourhejazy, P., Ying, K., & Lin, C. (2021). Unsupervised learning-based artificial bee colony for minimizing non-value-adding operations. *Applied Soft Computing*, 105, 107280.

Choy, K., Lee, W., Lau, H., Lu, D., & Lo, V. (2004). Design of an intelligent supplier relationship management system for new product development. *International Journal of Computer Integrated Manufacturing*, 17(8), 692–715.

Dhanalakshmi, R., DwarakaMai, C., Latha, B., & Vijayaraghavan, N. (2021). Ar and vr in manufacturing. In *Futuristic Trends in Intelligent Manufacturing* (pp. 171–183). UK/India: Springer.

Dogan, A. & Birant, D. (2020). Machine learning and data mining in manufacturing. *Expert Systems with Applications*, 166(2), 114060.

Dyke, H. V. & Parunak. (1990). Distributed ai and manufacturing control: Some issues and insights. *Decentralized AI*, 1, 81–99.

Epureanu, B., Li, X., Nassehi, A., & Koren, Y. (2020). Self-repair of smart manufacturing systems by deep reinforcement learning. *CIRP Annals*, 69(1), 421–424.

Fahle, S., Prinz, C., & Kuhlenkotter, B. (2020). Systematic review on machine learning (ml) methods for manufacturing processes — Identifying artificial intelligence (ai) methods for field application. *Procedia CIRP*, 93, 413–418.

Fesaghandis, G. S. & Pooya, A. (2018). Design of an artificial intelligence system for predicting success of new product development and selecting proper market-product strategy in the food industry. *International Food and Agribusiness Management Review*, 21(7), (1030-2019-593). DOI: 10.22004/ag.econ.284901.

Feyzioğlu, O. & Özkan, G. (2007). Evaluation of new product development projects using artificial intelligence and fuzzy logic. *International Journal of Economics and Management Engineering*, 1(11), 778–784.

Fischer, B., Rohringer, W., Panzer, N., & Hecker, S. (2017). Acoustic process control for laser material processing: Optical microphone as a novel "ear" for industrial manufacturing. *Laser Technik Journal*, 14(5), 21–25.

Floreano, D. & Wood, R. J. (2015). Science, technology, and the future of small autonomous drones. *Nature*, 521(7553), 460–466.

Furtado, C. (2020). IIoT-based inventory management. https://pdfs.semantic scholar.org/b25e/88ef416d7bca73424250737bcb657d973d01.pdf.

Ghosh, S., Gourisaria, M. K., Routaray, S., & Pandey, M. (2020). IIoT: A survey and review of theoretical concepts. In *Interoperability in IoT for Smart Systems* (pp. 223–236). Boca Raton: CRC Press. https://doi.org/10.1201/9781003055976.

Gilchrist, A. (2016). *Industry 4.0: The Industrial Internet of Things*. Berkeley/CA; and Thailand: Springer. https://doi.org/10.1007/978-1-4842-2047-4.

Guedes, M. D. V., Marques, M. S., Guedes, P. C., Vidor Contri, R., & Guerreiro, I. (2021). The use of electronic tongue and sensory panel on taste evaluation of pediatric medicines: A systematic review. *Pharmaceutical Development and Technology*, 26(2), 119–137.

Gujara, R., Adsuleb, A., Shrivastavac, A., Tewarib, A., & Ravala, H. K. (2020). Machine monitoring and data analytics for optimization of manufacturing and planning using IIoT. http://www.copen.ac.in/proceedings/copen10/copen/212.pdf.

Hassan, M., Huda, S., Sharmeen, S., Abawajy, J., & Fortino, G. (2020). An adaptive trust boundary protection for IIoT networks using deep-learning feature extraction based semi-supervised model. *IEEE Transactions on Industrial Informatics*, 17(4), 2860–2870. DOI: 10.1109/tii.2020.3015026.

Herbert, S. (2019). Why IIoT should make businesses rethink security. *Network Security*, 2019(7), 9–11.

Hou, B., Yu, W., & Li, B. (2017). Applications of artificial intelligence in intelligent manufacturing: A review. *Frontiers of Information Technology & Electronic Engineering*, 18(1), 86–97.

How Amazon is changing supply chain management. (2019). https://www.the balancesmb.com/how-amazon-is-changing-supply-chain-managementIIoT protocols, a. https://www.IoTcentral.io/blog/IIoT-protocols-for-the-beginners.

Huang, S. & Pan, Y. (2015). Automated visual inspection in the semiconductor industry: A survey. *Computers in Industry*, 66, 1–10.

IIoT protocols. (2015). https://www.automation.com/en-us/articles/2015-2/ IIoT-protocols.

Izagirre, U., Andonegui, I., Eciolaza, L., & Zurutuza, U. (2021). Towards manufacturing robotics accuracy degradation assessment: A vision-based data-driven implementation. *Robotics and Computer-Integrated Manufacturing*, 67, 102029.

Jaloudi, S. (2019). Communication protocols of an industrial Internet of Things environment: A comparative study. *Future Internet*, 11, 66.

Jayaram, A. (2016). Lean six sigma approaches for global supply chain management using industry 4.0 and IIoT. In *2016 2nd International Conference on Contemporary Computing and Informatics (IC3I)* (pp. 89–94). IEEE.

Jayaram, A. (2017). An IIoT quality global enterprise inventory management model for automation and demand forecasting based on cloud. In *2017 International Conference on Computing, Communication and Automation (ICCCA)* (pp. 1258–1263). IEEE.

Johnson, T. L., Fletcher, S. R., Baker, W., & Charles, R. L. (2019). How and why we need to capture tacit knowledge in manufacturing: Case studies of visual inspection. *Applied Ergonomics*, 74, 1–9.

Kan, C., Yang, H., & Kumara, S. (2018). Parallel computing and network analytics for fast industrial internet-of-things (IIoT) machine information processing and condition monitoring. *Journal of Manufacturing Systems*, 46, 282–293.

Kavitha, S. & Srinivasan, J. (2020). Iot-enabled electronic nose for fragrance measurement of textiles. *International Journal of Innovative Science and Research Technology*, 5, 1232–1235.

Kim, J., Lee, H., Jeong, S., & Ahn, S. (2021). Sound-based remote real-time multi-device operational monitoring system using a Convolutional Neural Network (CNN). *Journal of Manufacturing Systems*, 58, 431–441.

Kim, M., Park, K., Choi, S., Lee, J., & Kim, D. (2018). AR/VR based live manual for user-centric smart factory services. In *IFIP International Conference on Advances in Production Management Systems* (pp. 417–421). Cham: Springer, Springer. Cham. https://doi.org/10.1007/978-3-319-99707-0_52.

Klintong, N., Vadhanasindhu, P., & Thawesaengskulthai, N. (2012). Artificial intelligence and successful factors for selecting product innovation development. In *2012 Third International Conference on Intelligent Systems Modelling and Simulation* (pp. 397–402).

Kolokas, N., Vafeiadis, T., Ioannidis, D., & Tzovaras, D. (2020). A generic fault prognostics algorithm for manufacturing industries using unsupervised machine learning classifiers. *Simulation Modelling Practice and Theory*, 103, 102109.

Kotsiopoulos, T., Sarigiannidis, P., Ioannidis, D., & Tzovaras, D. (2021). Machine learning and deep learning in smart manufacturing: The smart grid paradigm. *Computer Science Review*, 40, 100341.

Krause, T. (2020). Ai-based discrete-event simulations for manufacturing schedule optimization. In *Proceedings of the 2020 4th International Conference on*

Algorithms, Computing and Systems, ICACS'20 (pp. 87–91). New York, NY, USA: Association for Computing Machinery.

Kwong, C. K., Jiang, H., & Luo, X. G. (2016). Ai-based methodology of integrating affective design, engineering, and marketing for defining design specifications of new products. *Engineering Applications of Artificial Intelligence*, 47, 49–60. Artificial Intelligence Techniques in Product Engineering.

Leber, J. (2020). General electric's San Ramon software center takes shape. *MIT Technology Review.* https://protect-eu.mimecast.com/s/2vzRCD8Jzuy BkELcWSB-L?domain=technologyreview.com" https://www.technology review.com/2012/11/28/114725/general-electric-pitches-an-industrial-internet/.

Lee, C. K. M. (2018). Development of an industrial internet of things (IIoT) based smart robotic warehouse management system. In *International Conference on Information Resources Management (CONFIRM)*. Association for Information Systems.

Lee, D. E., Hwang, I., Valente, C. M. O., Oliveira, J. F. G., & Dornfeld, D. A. (2006). Precision manufacturing process monitoring with acoustic emission. *International Journal of Machine Tools and Manufacture*, 46(2), 176–188. Springer. https://www.sciencedirect.com/science/article/pii/S08906955 05000933.

Levina, A., Kalyazina, S., Ershova, Al., & Schuur, P. C. (2020). *IIOT Within the Architecture of the Manufacturing Company*. New York, NY, USA: Association for Computing Machinery.

Li, X. & Zhang, C. (2020). Some new trends of intelligent simulation optimization and scheduling in intelligent manufacturing. *Service Oriented Computing and Applications*, 14, 149–151.

Li, X., Jia, X., Yang, Q., & Lee, J. (2020). Quality analysis in metal additive manufacturing with deep learning. *Journal of Intelligent Manufacturing*, 31(8), 2003–2017.

Liao, Y. Pierin Ramos, L. Saturno, M. Deschamps, F. Loures, E. & Luis Szejka, A. (2017). The role of interoperability in the fourth industrial revolution era. *IFAC-PapersOnLine*, 50(1), 12434–12439.

López, F. & Casillas, J. (2013). Artificial intelligence-based systems applied in industrial marketing: An historical overview, current and future insights. *Industrial Marketing Management*, 42(4), 489–495.

Lorenzo, S. F., Anorga, J., & Arrizabalaga, S. (2020). A survey of IIoT protocols: A measure of vulnerability risk analysis based on CVSS. *ACM Computing Surveys*, 53(2), 1–53.

Loske, M., Rothe, L., & Gertler, D. G. (2019). Context-aware authentication: State-of-the-art evaluation and adaption to the IIoT. *2019 IEEE 5th World Forum on Internet of Things (WF-IoT)*, 2019, (pp. 64–69). IEEE, Ireland. DOI: 10.1109/WF-IoT.2019.8767327.

Lu, Y., Witherell, P., & Jones, A. (2020). Standard connections for IIoT empowered smart manufacturing. *Manufacturing Letters*, 26, 17–20.

Madakam, S. & Uchiya, T. (2019). Industrial Internet of Things (IIoT): Principles, processes and protocols. In *The Internet of Things in the Industrial Sector* (pp. 35–53). Cham: Springer. https://doi.org/10.1007/978-3-030-24892-5_2.

Maghazei, O. & Netland, T. (2019). Drones in manufacturing: Exploring opportunities for research and practice. *Journal of Manufacturing Technology Management*, 31(6), 1237–1259.

Mar, N. S. S., Yarlagadda, P. K. D. V., & Fookes, C. (2011). Design and development of automatic visual inspection system for PCB manufacturing. *Robotics and Computer-Integrated Manufacturing*, 27(5), 949–962.

Meng, L., McWilliams, B., Jarosinski, W., Park, H., Jung, Y., Lee, J., & Zhang, J. (2020). Machine learning in additive manufacturing: A review. *JOM*, 72(6), 2363–2377.

Nozaki, N., Konno, E., Sato, M., Sakairi, M., Shibuya, T., Kanazawa, Y., & Georgescu, S. (2017). Application of artificial intelligence technology in product design. *Fujitsu Scientific and Technical Journal*, 53, 43–51.

Ntagios, M., Escobedo, P., & Dahiya, R. (2020). 3D printed robotic hand with embedded touch sensors. In *2020 IEEE International Conference on Flexible and Printable Sensors and Systems (FLEPS)* (pp. 1–4). IEEE.

Nti, I. K., Adekoya, A. F., Weyori, B. A., & Nyarko-Boateng, O. (2021). Applications of artificial intelligence in engineering and manufacturing: A systematic review. *Journal of Intelligent Manufacturing*, 1, 1–21. https://doi.org/10.1007/s10845-021-01771-6.

Oh, A. (2019). Development of a smart supply-chain management solution based on logistics standards utilizing artificial intelligence and the Internet of Things. *Journal of Information and Communication Convergence Engineering*, 17(3), 198–204.

Oliff, H., Liu, Y., Kumar, M., Williams, M., & Ryan, M. (2020). Reinforcement learning for facilitating human-robot-interaction in manufacturing. *Journal of Manufacturing Systems*, 56, 326–340.

Olivares, V., Cordova, F., Sepúlveda, J. M., & Derpich, I. (2015). Modeling internal logistics by using drones on the stage of assembly of products. *Procedia Computer Science*, 55, 1240–1249.

Paraschos, P. D., Koulinas, G. K., & KoulourIoTis, D. E. (2020). Reinforcement learning for combined production-maintenance and quality control of a manufacturing system with deterioration failures. *Journal of Manufacturing Systems*, 56, 470–483.

Prasad, R. & Choudhary, P. (2021). State-of-the-art of artificial intelligence. *Journal of Mobile Multimedia*, 17(1–3), 427–454.

Purcell, T. (May 14, 2019). R&D Director at Datel. https://www.iotforall.com/IoT-applications-manufacturing.

Rahman, T. M., Rahimi, A., Gupta, S., & Panat, R. (2016). Microscale additive manufacturing and modeling of interdigitated capacitive touch sensors. *Sensors and Actuators A: Physical*, 248, 94–103.

Rao, S. S., Nahm, A., Shi, Z., Deng, X., & Syamil, A. (1999). Artificial intelligence and expert systems applications in new product development — A survey. *Journal of Intelligent Manufacturing*, 10, 231–244.

Ren, J., Zhou, R., Farrow, M., Peiris, R., Alosi, T., Guenard, R., & Romero-Torres, S. (2020). Application of a kNN-based similarity method to biopharmaceutical manufacturing. *Biotechnology Progress*, 36(2), e2945.

Ren, L., Liu, Y., Wang, X. J., Lu, J., & Deen, M. J. (2020). Cloud-edge based lightweight¨ temporal convolutional networks for remaining useful life prediction in IIoT. *IEEE Internet of Things Journal*, 8(16), 12578–12587. DOI: 10.1109/jiot.2020.3008170.

Ren, L., Sun, Y., Wang, H., & Zhang, L. (2018). Prediction of bearing remaining useful life with deep convolution neural network. *IEEE Access*, 6, 13041–13049.

Sapot, B. (January 2, 2019). Should manufacturers ignore Industry 4.0. and IIOT? https://gomingo.io/industry-4-0-and-iiot-realistic-manufacturing-analytics/.

Saravanan, P. N., Seetharaman, A. S., & Sharma, A. (2019). Customer expectation from Industrial Internet of Things (IIoT). *Journal of Manufacturing Technology Management*, 30, 1161–1178.

Serizawa, Y. & Shomura, Y. (2019). A flexible acoustic sensing system and its application to IIoT — Manufacturing field site. In *Proceedings of the 5th International Conference on Sensors Engineering and Electronics Instrumentation Advances* (SEIA'2019) (pp. 18–23).

Shah, D., Wang, J., & He, Q. P. (2020). Feature engineering in big data analytics for IoT-enabled smart manufacturing — Comparison between deep learning and statistical learning. *Computers & Chemical Engineering*, 141, 106970.

Shevchik, S. A., Masinelli, G., Kenel, C., Leinenbach, C., & Wasmer, K. (2019). Deep learning for in situ and real-time quality monitoring in additive manufacturing using acoustic emission. *IEEE Transactions on Industrial Informatics*, 15(9), 5194–5203.

Siedler, C., Glatt, M., Weber, P., Ebert, A., & Aurich, J. C. (2021). Engineering changes in manufacturing systems supported by ar/vr collaboration. *Procedia CIRP*, 96, 307–312.

Stefano, F., Benzi, F., & Bassi, E. (2020). IIoT based efficiency optimization in logistics applications. *Asian Journal of Basic Science & Research*, 2(4), 59–73.

Talmoudi, S., Kanada, T., & Hirata, Y. (2019). An IoT-based failure prediction solution using machine sound data. In *2019 IEEE/SICE International Symposium on System Integration (SII)* (pp. 227–232). IEEE.

Ungurean, I., Gaitan, N. C., & Gaitan, V. G. (2016). A middleware based architecture for the industrial Internet of Things. *KSII Transactions on Internet & Information Systems*, 10(7), 2874–2891.

Wang, J., Erkoyuncu, J., & Roy, R. (2018). A conceptual design for smell based augmented reality: Case study in maintenance diagnosis. *Procedia CIRP*, 78, 109–114.

Wocker, M., Betz, N. K., Feuersanger, C., Lindworsky, A., & Deuse, J. (2020). Unsupervised learning for opportunistic maintenance optimization in flexible manufacturing systems. *Procedia CIRP*, 93, 1025–1030.

Xia, K., Sacco, C., Kirkpatrick, M., Saidy, C., Nguyen, L., Kircaliali, A., & Harik, R. (2021). A digital twin to train deep reinforcement learning agent for smart manufacturing plants: Environment, interfaces and intelligence. *Journal of Manufacturing Systems*, 58, 210–230.

Yang, C., Shen, W., & Wang, X. (2016). Applications of Internet Of Things in manufacturing. In *2016 IEEE 20th International Conference on Computer Supported Cooperative Work in Design (CSCWD)* (pp. 670–675).

Yu, J., Li, X., Lu, W. F., & Sun, Y. (2020). Fm-based supervised learning for categorical data classification in manufacturing process. In *2020 15th IEEE Conference on Industrial Electronics and Applications (ICIEA)* (pp. 1438–1443). IEEE.

Zelbst, P. J., Green, K. W., Sower, V. E., & Bond, P. L. (2019). The impact of rfid, IIoT, and blockchain technologies on supply chain transparency. *Journal of Manufacturing Technology Management*, 31(3), 441–457.

Zhang, B., Anderljung, M. S., Kahn, L., Dreksler, N., Horowitz, M. C., & Dafoe, A. (2021). Ethics and governance of artificial intelligence: Evidence from a survey of machine learning researchers. arXiv preprint arXiv:2105.02117.

Zhang, Y., Guo, Z., Lv, J., & Liu, Y. (2018). A framework for smart productionlogistics systems based on cps and industrial IoT. *IEEE Transactions on Industrial Informatics*, 14(9), 4019–4032.

Zhao, Z., Lin, P., Shen, L., Zhang, M., & Huang, G. Q. (2020). Iot edge computing enabled collaborative tracking system for manufacturing resources in industrial park. *Advanced Engineering Informatics*, 43, 101044.

Zheng, H., Paiva, A. R., & Gurciullo, C. S. (2020). Advancing from predictive maintenance to intelligent maintenance with ai and IIoT. arXiv preprint arXiv:2009.00351.

Zhou, G., Zhang, C., Li, Z., Ding, K., & Wang, C. (2020). Knowledge-driven digital twin manufacturing cell towards intelligent manufacturing. *International Journal of Production Research*, 58(4), 1034–1051.

Chapter 8

Fighting Food Waste: How Can Artificial Intelligence and Analytics Help?

Lohithaksha M. Maiyar[*,†,¶]**, Ramakrishnan Ramanathan**[‡,||]**,
Shanta Lakshmi Belavadi Nagaraja Swamy**[*,**]
and Usha Ramanathan[§,††]

[*]*Business & Management Research Institute, University of Bedfordshire
Business School, Luton, UK*
[†]*Department of Entrepreneurship and Management, Indian Institute of
Technology Hyderabad, Kandi, Telangana, India*
[‡]*Essex Business School, University of Essex,
Southend-on-Sea, UK*
[§]*Ajman University, Ajman, UAE and Nottingham Trent University,
Nottingham, UK*
[¶]*l.maiyar@em.iith.ac.in*
[||]*r.ramanathan@essex.ac.uk*
[**]*lak780@yahoo.com*
[††]*usha.ramanathan@ntu.ac.uk*

The application domains of Artificial Intelligence (AI) and Analytics are ever increasing, specifically in food production globally. The availability of huge amounts of data, machine-to-machine communications, cloud storage, blockchain and the emergence of newer methods of data analytics have helped to kindle renewed interest in AI developments for the European food sector. Some experiences from a large European project, called the REAMIT project, will be used in this chapter to discover the possibilities of using AI to identify causes, patterns, and solutions to the food wastage crisis. Approximately one-third of food produced gets wasted in both developed and developing nations. The REAMIT project uses Internet of Things (IoT) sensors to sense food quality. This chapter will uncover some promising potentials of using AI techniques with sensor data. The various criteria that influence the selection of the right AI tool under various scenarios of the IoT system implementation are captured through a hierarchical multi-criteria framework. This chapter also proposes an innovative analytics framework for food waste reduction which can be adapted widely by individuals and practitioners.

1. Introduction

The ramifications of digital technology across food supply chains and the extended use of AI and innovation analytics to address the problems beyond profit objectives, such as reduction of fresh food wastage and the associated carbon emissions, have paved the way for new research paradigms and sustainable industrial practices in the arena. The twelfth sustainable development goal (United Nations: Department of Economic and Social Affairs Sustainable Development, 2021) is to reduce the food waste by propelling policies to achieve responsible consumption and production patterns. In 2016, 13.8% of food was wasted across supply chains globally, while the European and North American regions accounted for 15.7% wastage across food supply chains (SDG Indicators, 2021). In line with these concerns, the REAMIT project (REAMIT — Improving resource efficiency of agribusiness supply chains by minimizing waste using Big Data and Internet of Things sensors | Interreg NWE, 2021) aims to reduce the fresh food wastage across Northwest Europe by at least 10% with the use of IoT sensors, big data analytics, and AI. The key is to use IoT sensors to track food quality while the food is traveling along the supply chains. The sensor information is regularly sent to the cloud so that data analytics teams and special AI-based algorithms can be used to track the quality for real-time monitoring.

The use of IoT sensors is not always necessary as some food items (e.g., rice and other grains) are not very sensitive to temperature/

humidity variations. They can be transported without any tracking and are not prone to food loss due to temperature fluctuations. However, fresh produce (e.g., chilled food items such as fruits and vegetables and frozen food items such as frozen fruits and vegetables) that is very sensitive to temperature fluctuations needs tighter controls using IoT sensors. In these cases, many logistics providers use temperature-controlled trucks for transportations. However, at least two variations of such trucks can be observed in practice. Some trucks have temperature control but do not have arrangements for the temperature (and other) sensors to send out their observations as the sensors are not IoT sensors. In these trucks, the variations in temperature can be logged in a data logger that can be read at the end of the journey. While data loggers are useful to identify potential faults in the temperature control systems, it might become too late to help in maintaining food quality and avoiding food waste as food may have already been spoilt in case of any malfunction before the truck reaches its destination. To avoid such issues, the REAMIT project proposes to use IoT sensors that not only record temperature in the trucks but also transmit these observations to the cloud as they are connected to the internet via a gateway. Within this context, this chapter distinguishes three different approaches for tracking food quality in the supply chain: (1) temperature monitoring but no control, (2) temperature monitoring with control, and (3) with no temperature monitoring and control. Further, the scope for the use of big data and AI tools for reducing food waste is reviewed with special emphasis on these three different approaches.

The literature review is synthesized in the form of a multi-criteria framework for selecting an appropriate AI tool for reducing food waste. In addition, in line with the theme of this book, this chapter provides an innovative analytics framework for food waste reduction.

The sections of this chapter are organized in the following way. Section 2 reviews some important definitions linked to food waste to ease the understanding of readers. Section 3 describes the literature review conducted as part of the study to investigate food waste reduction possibilities under different tracking conditions. Section 4 presents a multi-criteria decision-making hierarchical framework for selecting the appropriate AI tool and an analytics implementation approach for reducing food waste. Section 5 finally describes few examples in which the IoT technology was implemented and identifies scope for further research and case studies.

2. Some Definitions on Food Waste

Before delving into how digital food technology and AI can help reduce food waste, the correct understanding of terms such as food wastage, food quality, and the link between them is important to address the context of reducing food wastage through the use of AI and big data. According to Tavill (2020), food waste is defined as the "inedible by-products of food production practices". These inedible parts include farming residues like corn silage and stems, and are normally managed at the farm level. Inedible parts also include components removed at the time of preparation or processing at home, at a food service venue, or at a factory. The same author identifies the definition of wasted food differently as "uneaten edible food, which is largely generated at the consumer level either at or away from home". According to the Food and Agriculture Organization (FAO), food loss or food waste is defined as "food that is not eaten". However, for the purpose of this chapter, food waste specifically refers to the resulting loss of fresh food because of it becoming inedible while the food is being transported or is at the risk of becoming inedible while it reaches the supermarket or the end consumer.

It is important to highlight that any food waste or wasted food that may arise after the food reaches the customer is considered as out of the scope of this research. Table 1 highlights similar but slightly different definitions of food waste put forth by numerous authors under a variety of contexts in the fresh food supply chains.

Balaji & Arshinder (2016) have identified 16 causes of food wastage in perishable food supply chains for the Indian context out of which lack of refrigerated carriers, poor logistics infrastructure, poor logistics network design, lack of integrated IT systems, lack of coordination among players and lack of traceability systems are some of the important issues which are globally relevant. Attributed to one or more of the aforementioned causes, the quality of the fresh food eventually falls short of edible standards. Akkerman *et al.* (2010) highlighted the importance of maintaining food quality using temperature-controlled distribution and tracking systems in order to reduce fresh food wastage. According to (Grunert, 2005), food quality refers to the maintenance of physical properties of food products within their appropriate thresholds, and also the way the product is perceived by the final consumer, which could be microbial aspects, texture, or flavor of the product delivered. Some additional examples of food quality metrics include total bacterial count, total specific

Table 1.   Definitions, sources, and causes of food wastage.

Source	Definition	Source of wastage	Cause

Tavill (2020)	• Wasted food is defined as
	  – "uneaten edible food and is largely generated at the consumer level either at or away from home".
	  – "edible" materials that are not eaten due to a choice or event.
	  – a component of packaging.
	• Food waste
	  – is defined as inedible by-products of food production practices and this is on the rise due to economic, social/ethical, and environmental factors.
	  – refers to the "inedible" organic materials remaining after food production, processing, or preparation. These inedible parts include farming residues like corn silage and stems that are normally managed at the farm level. Inedible parts also include components removed at the time of preparation or processing at home, at a food service venue, or at a factory.
		Consumer level, household level.	Food storage, preservation and processing techniques, packaging technologies.

Gaiani *et al.* (2018)	• Household food waste is defined as the food waste occurring between acquisition and food preparation, food preparation and food serving, and after food serving (plate waste).	Consumption stage of the food supply chain.	Consumers' attitude to waste food.
Principato *et al.* (2015)	• Classifications for household food waste: avoidable and possibly avoidable waste that is the "edible" food thrown away; and unavoidable food waste which is waste derived from food preparation that is not, and was not edible, such as bones, shells, and skins.	Pre-shopping and consumption phases.	Consumer behavior concerning food waste and planning shopping.

bacterial count, food freshness index, color of the food skin, odour of the food product and the physical structure (damaged or intact) of the food product.

The first step towards capturing food waste is to detect food quality directly or indirectly by deploying several traditional sensors as well as advanced technologies such as image recognition, Raman spectroscopy, IoT sensors, liquid bacterial detection and 3D fluorescence in combination with AI and big data technologies. This paradigm shift in the application of analytics and AI to improve the holistic performance of a system though the innovation process is termed as innovation analytics (Balaji & Arshinder, 2016). Reduction of food wastage requires a similar effort in the fresh food supply chain which can also be referred to as food supply chain innovation analytics. In the next section, a critical review of articles that focus on different categories of temperature tracking and control environments is conducted by revisiting the literature on fresh food supply chains. Furthermore, the study presents an analytics framework for reducing food waste.

3. Food Waste Reduction Under Different Tracking Conditions

The literature review for this study is conducted by collecting recently published articles from scientific databases such as Web of Science, EBSCO, Science Direct, and Google Scholar. A general outline on the topic was gathered by conducting a search using keywords such as "artificial intelligence in reducing food waste", "ANN", and "neural network" in combination with "for fresh food supply chain", "temperature control", "temperature monitoring", "food quality control", and "food quality monitoring". Papers which focused on household waste, production waste, or any other tiers of supply chain other than during the distribution phase were explicitly excluded from the search. Papers published in English and with accessibility to read the full paper were considered. Existing literature on food wastage reduction using digital and AI technologies has considered the importance of the different dimensions such as humidity, temperature, and packaging for improving the shelf life of the product and for simultaneously reducing the food wastage. This chapter mainly emphasizes studies that capture distribution systems with monitoring and controlling of the temperature for reducing food wastage.

The articles reviewed are segregated into three categories as shown in Table 2: (1) systems with temperature monitoring and no control, (2) systems with both temperature monitoring and control, and (3) systems with no temperature monitoring and control.

3.1. *Temperature monitoring and no control*

A new temperature estimation procedure has been evaluated by Akkerman *et al.*, (2010) in which thermal imaging is used to predict surface temperature over a pallet of apples while comparing packaging of plastic boxes and cardboard boxes. This temperature data are then introduced as an input to the artificial neural network (ANN) software to estimate the temperature across the entire pallet. Using thermal imaging alone would only allow the surface temperature to be obtained, whereas their method helped to also measure the temperatures across the freight with high levels of accuracy. Grunert (2005) has estimated the accuracy of the temperature distribution inside a food pallet based on prediction from a heat transfer model using measurements at a single position inside the pallet. Furthermore, the heat transfer coefficient is estimated from the simulated measurements and then used with the heat transfer model to map the entire temperature distribution inside the pallet. Furthermore, Tavill (2020) proposed to leverage the theoretical foundation and generalization ability of the physical heat transfer model to develop a flexible neural net framework which can predict temperatures in real-time. In this regard, simulations show that the ANN can predict the temperature distribution inside a pallet with an average error below 0.5 K in a one-sensor-per-pallet scenario when the sensor is properly located inside the pallet. A framework for quality-driven distribution for perishables is developed by Gaiani *et al.* (2018) which integrates intelligent container within the IoT system to provide information for reducing food waste. Principato *et al.* (2015) have validated an RFID smart tag developed for real-time traceability and cold chain monitoring for food applications. Kakatkar *et al.* (2020) have discussed a sensor-based method for real-time quality monitoring which is found to be superior than the traditional visual assessment method. In this, product metabolism was assessed   based on temperature changes and Euclidean distance cost. Badia-Melis *et al.* (2016b) proposed new solutions using new RFID temperature monitoring technology which empowers the fresh fruits and vegetables industry to significantly reduce waste. Mercier *et al.* (2017) have demonstrated the combination of weather prediction software and mathematical modeling of heat transfer to quantify the performance of insulated boxes exposed to real-world conditions.

Mercier & Uysal (2018) have compared Newton's law of cooling with ANN and auto-regressive moving average models with respect to deviation prediction, prediction error and execution time. Chen & Shaw (2011) have evaluated an exponentially weighted moving average (EWMA) control chart and ANN technologies in order to monitor collected temperature data in the context of cold chain management. Dittmer *et al.* (2012) have used miniaturized RFID temperature loggers to analyze the

Table 2.   Classification of reviewed literature into different categories of temperature monitoring and controlling abilities.

Literature source	Systems with no temperature control and temp monitoring	Systems with temperature monitoring but no control	Systems with temperature monitoring and control
Aung & Chang (2014)	x	√	x
Do Nascimento Nunes *et al.* (2014)	x	√	x
East *et al.* (2009)	x	x	√
Tavill (2020)	x	√	x
Konovalenko *et al.* (2021)	x	x	√
Jedermann *et al.* (2009)	x	x	√
Gaiani *et al.* (2018)	x	√	x
Grunert (2005)	x	√	x
Kakatkar *et al.* (2020)	x	√	x
Lang *et al.* (2011)	x	√	x
Badia-Melis *et al.* (2016b)	x	x	√
Badia-Melis *et al.* (2016a)	x	√	x
Raab *et al.* (2011)	x	√	x
Kim *et al.* (2015)	x	x	√
Badia-Melis *et al.*, (2016b)	x	x	√
Abad *et al.* (2009)	x	√	x
Boquete *et al.* (2010)	√	x	x

amount of local deviations, detect temperature gradients, and estimate the minimum number of sensors that are necessary for reliable monitoring inside a truck or container. In their case, the authors have discussed the use of sensors or IoT technology to analyze the sensor data and partially monitor through an automated system. Abad *et al.* (2009) argued that the intelligent container is considered as enabling tool for the new logistic paradigm of FEFO (first expire first out) taking into account the remaining shelf life and increasing the probability of delivering better quality goods to the customer, with lesser losses during transport and a reduced carbon footprint.

The selection of the right AI tool for implementation of an AI-enabled temperature monitoring system is associated with some additional cost to the food transporter as compared to a case of transport with no temperature monitoring and control. Some of the important factors that contribute to the overall cost incurred for its implementation are weather conditions, number and type of sensors used, number of control zones in operation, type of IoT technology used and distance and unit cost of transportation (Jedermann *et al.*, 2009; Accorsi *et al.*, 2017). Based on the type of technology used to maintain the quality of food there could be other equipment and operational costs associated with reading and analyzing the data collected. In addition to cost, few other criteria also play a role in the selection of the right AI tool such as quality aspects and category of external environment (ambient, chilled frozen). Quality aspects can either refer to level of contamination of food or to maintaining associated standards of the technologies adopted for monitoring the variables.

3.2. *Temperature monitoring and control*

A low-cost and highly versatile temperature monitoring system proposed by Jedermann *et al.* (2009) is applicable to all phases of wine production, from grape cultivation through to delivery of bottled wine to the end customer. Here, monitoring is performed by a purpose-built electronic system comprising a digital memory that stores temperature data and a ZigBee communication system that transmits data to a control center for processing and display. Moreover, with minimum modification, other variables of interest (pH, humidity, etc.) could also be monitored and the system could be applied to other similar sectors, such as olive oil production. East *et al.* (2009) conducted a literature review of existing temperature monitoring systems and solutions used for optimal temperature monitoring in meat supply chains. Konovalenko *et al.* (2021) have introduced a concept for measuring quality called Freshness Gauge and an algorithm that adjusts proper temperature and humidity levels by reflecting the quality change of eatable products in refrigerated storage. Do Nascimento Nunes *et al.* (2014) have compared three distinct estimation algorithms: (i) the capacitive heat transfer method, (ii) the Kriging algorithm and (iii) ANN. According to Do Nascimento Nunes *et al.* (2014), algorithmic temperature mapping and prediction in transportation containers is a very effective tool in developing cold chain strategies with better monitoring to significantly reduce product losses and reducing waste. Badia-Melis *et al.* (2016b) have proposed a model which was applied to an illustrative case study of a cold chain for cherries and demonstrates the influence of the weather conditions on the energy costs for the products' refrigeration in vehicles during transportation and at the warehouse during storage. Kim *et al.* (2015) have presented an improved approach to cold chain management (CCM) based on the real-time monitoring of perishable cargo using RFID-based sensing techniques, combined with the modeling of current and future in-cargo temperatures using the available sensed data.

The AI-enabled implementation of a temperature monitoring and control system is associated with some additional costs as compared to the previous case such as cost incurred for implementation of real-time automated closed-loop control, additional labor effort costs, labor health insurance costs and real-time computational costs. The need to measure and retain quality of food for different external environments would remain the same for this case and will be governed by the same set of criteria and sub-criteria as defined for the earlier case.

3.3. *No temperature control and monitoring*

A majority of the research studies related to cold chain distribution fall into this category where the authors capture distribution of fresh food items in manually temperature-controlled pallets or vans which however do not have the ability to act upon any unexpected temperature variation around the food item due to lack of a real-time monitoring and controlling digital system. Boquete *et al.* (2010) identified the issues and challenges with the existing temperature-controlled cold chain logistics systems and emphasized the need for efficient monitoring and control to maintain food quality. The cost incurred for implementation of a system with no monitoring and control would be much lesser than the cases presented earlier. In the former case, costs are only incurred to maintain the refrigerated systems in their cold vans or transport systems. This system will however be associated with a higher risk of food spoilage. Boquete *et al.* (2010) presented several cases of temperature failures while transporting various types of food products such as fruits, vegetables, and meat across different countries indicating the dire need for developing new models for the implementation of real-time monitoring and control of fresh food products.

4. A Multi-Criteria Decision-Making Framework and Analytics Implementation Approach

Real-time monitoring of data from IoT sensors results in huge amounts of data. A single truck with 20 sensors that send observations on temperature every 5 minutes will produce 240 observations every hour, 5760 observations every day, and more than 2 million observations in a year. Traditional analysis tools may not be appropriate to deal with such voluminous data and therefore there is a need for dealing with big data-based analytics approaches (Ramanthan *et al.* 2017). AI-based tools which have faster computational ability will gain more importance to make sense of such data. A detailed discussion of various AI-based tools is beyond the scope of this chapter. Interested readers can refer to standard analytics materials, (e.g., East *et al.*, 2009). A number of AI-based tools are available, but it is difficult to decide on the most appropriate tools for an application. Based on a literature review conducted in this chapter, a multi-criteria hierarchical framework for choosing the right AI tool for food quality monitoring is provided in Figure 1. The figure lists a number of AI tools. A closed-loop innovative analytics implementation approach is proposed for reducing waste in Section 4.1.

As Figure 1 shows, a multi-criteria hierarchical framework is proposed in which the first level represents the goal to arrive at a unique selection of different AI tools which have been used for optimal monitoring of fresh food. Further classification of AI tools used to track quality or reduce food waste is made based on three important criteria which are cost, quality, and external environment in the second level of the hierarchy. The proposed hierarchy comprises of at least four levels under each

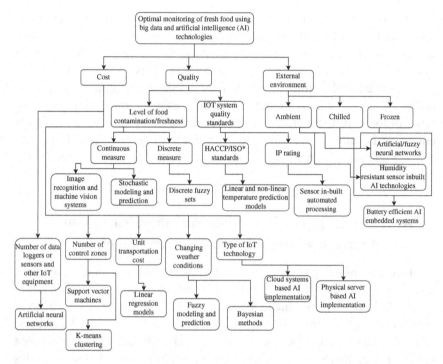

Figure 1.   A multi-criteria hierarchical framework to assist the selection of the right AI procedure for food quality monitoring.

Note: *HACCP — Hazard Analysis and Critical Control Point, ISO — International Organization for Standardization.

important category with quality criteria of the hierarchical structure having five levels. Here, cost and quality are the two important criteria in deciding the data requirements (more frequent or less frequent data generation) and level of control needed (monitoring or no monitoring). The nature of the produce to be transported will be shaped by the external environment. The third level of hierarchy enlists the various contexts where AI can be adopted under each criteria presented in the second level. For example, the cost of combined implementation of IoT and AI technology depends on several factors such as number of sensors and IoT equipment used. Accurate prediction requires higher amounts of data, whereas increasing the number of sensors without restriction is definitely not cost effective due to the high individual cost of sensors. Similarly, the levels of many other factors such as the number of temperature-controlled zones required, unit transportation cost, changing weather conditions, and type of AI technology used should be carefully chosen.

The implementation of AI and IoT technology for quality concerns can be classified under two aspects, technologies for measurement and tracking of food quality and technologies for maintaining IoT and food quality standards. For example, technologies such as image recognition, machine vision systems, stochastic modeling, and discrete fuzzy systems are commonly used for predicting food freshness, whereas linear/nonlinear temperature prediction models are commonly used for adhering to HACCP regulations, and sensors inbuilt with automated processing capability should have the Ingress Protection (IP) rating compliant with international IoT implementation standards.

The type of external or surrounding environment around the food product also significantly influences choice of the AI tool. For example, frozen food environments require more battery-efficient AI-embedded systems. The framework in Figure 1 is likely to be adapted and improvised as and when more AI tools are considered, and more criteria are added to the framework.

4.1. *Innovative analytics approach to reduce food waste*

The innovative analytics approach has been derived by experiences from technology demonstrations currently underway in the REAMIT project. Midway transport management of the food item is possible only through an efficient data analytics framework to analyze and send feedback to appropriate players to quickly act upon either retaining the quality or diverting the food item as and when feasible to nearby alternative destinations based on the spoilage level and degree of the risk of spoilage of the food item at any instantaneous moment of the transport. The proposed innovative analytics framework in Figure 2 is designed to improve the present condition of food waste with the help of an integrated digital IoT system. The framework aims to capture and analyze real-time information for improvising the operational condition by taking corrective action and consequently reducing the amount of food otherwise wasted, thereby forming a closed-loop semi-automated control system for reducing food waste. The idea of having a semi-automated control system is inspired from the fact that a fully automated temperature-controlled system is associated with heavy implementation cost.

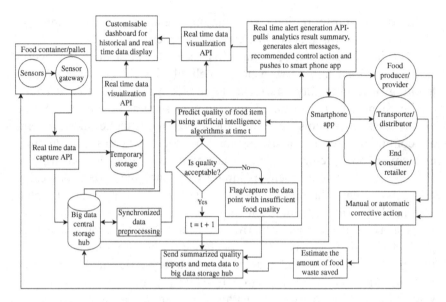

Figure 2.   Proposed innovative analytics framework for food waste reduction.

The framework proposed in Figure 2 represents the flow of data from the sensor source through a connected set of automated and manual processes required for completing a closed-loop real-time IoT innovative decision support. The data from the sensors will be initially captured with the help of wireless technology and sent to a temporary cloud storage, which will then be subsequently sent to a permanent physical big data storage. The purpose of having a physical storage is to provide a central storage facility for all actors of the real-time decision support system and to avoid third-party data transfers. The data are pulled by a data analytics module either located locally or in an external data analytics client machine to be analyzed by the analytics person. A time synchronized data pre-processing procedure will be needed to filter the data being received from the sensors to the format required by the data analytics module. The data analytics module will contain the appropriate AI technique to predict food quality and other parameters of the real-time IoT system such as the number of sensors and classification of control zones to name a few, which would be essential to improve the IoT system performance. The outcomes and summarized quality reports will be routed back to the server which will then be pulled by three modules, the alert generation API, the smart phone application, and the visualization API, to generate alerts with respect to food quality levels and communicate them to different actors of the supply chain. The smart phone application will recommend a corrective action for different abnormal situations observed after analysis based on a historical corrective action information database which, based on stakeholder's choice, can be actioned either by the nearest actor or by an automated system. The decision to opt in or opt out of having an automated system feedback control or a manual one would directly depend on the rate of returns on the respective investment options. The proposed real-time analytics approach is robust and applicable across different food supply chain configurations because of the capability of the system to receive and process data from a wide variety (moving, stationery, and weather variations) and forms (numerical, categorical, pictorial, spectral, time series, and non-time series) of data sources. This framework can be tested by any food manufacturer/supplier/distributor with a basic investment in sensor and data storage facilities. Based on initial data analysis, the success can be readily predicted before a huge investment in AI and data analytics.

5. Scope for Further Research Involving Technology Demonstration and Case Studies

AI and Data analytics are widely used in European countries in the food sector. AI is also used in production facilities to replace various types of manual labor such as picking fruits and vegetables of specified size and quality. Also, some AI technologies are helping in sowing and harvesting some large agricultural fields given that they are leveled and structured.

It is worthwhile to note some examples of the use of AI techniques and real-time analytics in reducing food waste; for example, Jagtap *et al.* (2019) have shown the ability to determine the potential reasons for potato waste generation using Convolutional Neural Network (CNN) achieving an accuracy of 99.79% on a small set of samples in their training test. They were able to measure the total amount of waste at real time along with finding the reasons for waste generation within the potato processing industry. This is by using image processing and load cell technologies where the images were captured through a purposefully positioned camera and were then processed to detect the unusable, damaged potatoes, and a digital load cell was used to measure their weight. Subsequently, deep-learning architecture using CNN was utilized to determine a potential reason for the potato waste generation and eventually aid in the implementation of corrective actions (Chen & Shaw, 2011). Aung & Chang (2014) showed the use of a thermal imaging technique to predict the surface temperature over a pallet of apples while comparing packaging (plastic boxes and cardboard boxes). The temperature data were then introduced as an input in ANN software to estimate the temperature across the entire pallet. This chapter demonstrated the possibility of temperature monitoring by ANN using thermal imaging technology (Aung & Chang, 2014). Some of these examples show the potential use of AI techniques and real-time analytics to be effective in reducing food waste. In this line of thought, REAMIT currently is at the point of collecting data which are intended to be utilized for monitoring and reducing food waste eventually.

The multi-criteria hierarchical framework and the real-time innovative analytics implementation approach proposed in this work are being implemented across multiple pilot tests (also called technology demonstration) of the REAMIT project across several regions of Europe and the United Kingdom. For example, a pilot test in the Netherlands firstly aims to develop a customized cooling profile for every crate of food transported for the last-mile delivery and secondly aims to map the customer complaints with acceleration profiles to ensure the food is handled properly and reaches the customer in the right shape and quality. The data from this pilot test are in numerical format with several thousands of rows to capture information such as crate temperature, weather temperature, humidity, and vibration for every time instant with inter-reading frequency of 5 mins. In another case, a pilot test in France uses Raman spectroscopy to generate spectral data to analyze meat quality on samples of meat. This data are non-time series and contain thousands of rows that correspond to spectral information measured instantaneously. Similarly, there are more pilots planned for execution across the United Kingdom, Ireland, and France under the context of the REAMIT project. The appropriate choice of AI technology to predict food quality and the selection of performance improvement variables in the aforementioned pilot tests will be made using the multi-criteria hierarchical framework proposed in this study.

The novel multi-criteria AI tool hierarchical framework and innovative analytics implementation approach proposed in this work leads to several new avenues for future research. For example to identify the most influential AI tool for reducing foodwaste and for further investigating the interdependencies between the various AI tools identified in the proposed hierarchical classification using procedures such as analytical hierarchy process, analytical network process and other hybrid fuzzy multi-criteria decision-making approaches. The proposed analytics approach holds further scope for improvement by integrating it with block chain concepts for improving data security and transparency.

Acknowledgment

The research in this chapter is based on work done on the REAMIT project funded by Interreg North-West Europe (Project number NWE831).

References

Abad, E., Palacio, F., Nuin, M., Zárate, A. G. de, Juarros, A., Gómez, J. M., *et al.* (2009). RFID smart tag for traceability and cold chain monitoring of foods: Demonstration in an intercontinental fresh fish logistic chain. *Journal of Food Engineering*, 93(4), 394–399.

Accorsi, R., Gallo, A., & Manzini, R. (2017). A climate driven decision-support model for the distribution of perishable products. *Journal of Cleaner Production*, 165, 917. https://doi.org/10.1016/j.jclepro.2017.07.170.

Akkerman, R., Farahani, P., & Grunow, M. (2010). Quality, safety and sustainability in food distribution: A review of quantitative operations management approaches and challenges. *OR Spectrum*, 32(4), 863–904.

Aung, M. M. & Chang, Y. S. (2014). Temperature management for the quality assurance of a perishable food supply chain. *Food Control*, 40(1), 198–207.

Badia-Melis, R., Mc Carthy, U., & Uysal, I. (2016a). Data estimation methods for predicting temperatures of fruit in refrigerated containers. *Biosystems Engineering*, 151, 261–272.

Badia-Melis, R., Qian, J. P., Fan, B. L., Hoyos-Echevarria, P., Ruiz-García, L., & Yang, X. T. (2016b). Artificial neural networks and thermal image for temperature prediction in apples. *Food Bioprocess Technology*, 9(7), 1089–1099.

Balaji, M. & Arshinder, K. (2016). Modeling the causes of food wastage in Indian perishable food supply chain. *Resources, Conservation & Recycling*, 114, 153.

Boquete, L., Cambralla, R., Rodrguez-Ascariz, J. M., Miguel-Jimnez, J. M., Cantos-Frontela, J. J., & Dongil, J. (2010). Portable system for temperature monitoring in all phases of wine production. *ISA Transactions*, 49, 3.

Chen, K. Y. & Shaw, Y. C. (2011 Jan 1). Applying back propagation network to cold chain temperature monitoring. *Advanced Engineering Informatics,* 25(1), 11–22.

Dittmer, P., Veigt, M., Scholz-Reiter, B., Heidmann, N., & Paul, S. (2012). The intelligent container as a part of the Internet of Things: A framework for quality-driven distribution for perishables. In *Proceedings — 2012 IEEE International Conference on Cyber Technology in Automation and Intelligent Systems, CYBER 2012, 209–214: Control.*

Do Nascimento Nunes, M. C., Nicometo, M., Emond, J. P., Melis, R. B., & Uysal, I. (2014). Improvement in fresh fruit and vegetable logistics quality: Berry logistics field studies. *Philosophical Transactions of the Royal Society A,* 372(2017), 1–19.

East, A., Smale, N., & Kang, S. (2009). A method for quantitative risk assessment of temperature control in insulated boxes. *International Journal of Refrigeration,* 32, 6.

Gaiani, S., Caldeira, S., Adorno, V., Segre, A., & Vittuari, M. (2018). Food wasters: Profiling consumers' attitude to waste food in Italy. *Waste Management,* 72, 17.

Grunert, K. G. (2005). Food quality and safety: Consumer perception and demand. *European Review of Agricultural Economics,* 32(3), 369–391.

Jagtap, S., Bhatt, C., Thik, J., & Rahimifard, S. (2019). Monitoring potato waste in food manufacturing using image processing and Internet of Things approach. *Sustainability,* 11(11), 3173.

Jedermann, R., Ruiz-Garcia, L., & Lang, W. (2009). Spatial temperature profiling by semi-passive RFID loggers for perishable food transportation. *Computers and Electronics in Agriculture,* 65(2), 145–154.

Kakatkar, C., Bilgram, V., & Füller, J. (2020). Innovation analytics: Leveraging artificial intelligence in the innovation process. *Business Horizon,* 63(2), 171–181.

Kim, W. R., Aung, M. M., Chang, Y. S., & Makatsoris, C. (2015). Freshness Gauge based cold storage management: A method for adjusting temperature and humidity levels for food quality. *Food Control,* 47, 510–519.

Konovalenko, I., Ludwig, A., & Leopold, H. (2021). Real-time temperature prediction in a cold supply chain based on Newtons law of cooling. *Decision Support System,* 141(11345), 1.

Lang, W., Jedermann, R., Mrugala, D., Jabbari, A., Krieg-Brückner, B., & Schill, K. (2011). The "intelligent container" — A cognitive sensor network for transport management. *IEEE Sensors Journal,* 11(3), 688–698.

Mercier, S. & Uysal, I. (2018). Neural network models for predicting perishable food temperatures along the supply chain. *Biosystems Engineering,* 171, 91.

Mercier, S., Marcos, B., & Uysal, I. (2017). Identification of the best temperature measurement position inside a food pallet for the prediction of its temperature distribution. *International Journal of Refrigeration*, 76, 147.

Principato, L., Secondi, L., & Pratesi, C. A. (2015). Reducing food waste: An investigation on the behavior of Italian youths. *British Food Journal*, 117(2), 731–748.

Raab, V., Petersen, B., & Kreyenschmidt, J. (2011). Temperature monitoring in meat supply chains. *British Food Journal*, 113, 1267–1289.

Ramanathan, R., Mathirajan, M., & Ravindran, A. R., (eds.) (2017). *Big Data Analytics Using Multiple Criteria Decision Making Models*. Taylor & Francis, Florida, USA: CRC Press.

REAMIT — Improving Resource Efficiency of Agribusiness supply chains by Minimising waste using Big Data and Internet of Things sensors | Interreg NWE. https://www.nweurope.eu/projects/project-search/reamit-improving-resource-efficiency-of-agribusiness-supply-chains-by-minimising-waste-using-big-data-and-internet-of-things-sensors/.

SDG Indicators. https://unstats.un.org/sdgs/report/2020/.

Tavill, G. (2020). Industry challenges and approaches to food waste. *Physiology & Behavior*, 223, 112993.

United Nations: Department of Economic and Social Affairs Sustainable Development. https://sdgs.un.org/#goal_section.

Chapter 9

Intelligent Traffic Solutions (Role of Machine Learning and Machine Reasoning)

**Shivani Agarwal*, Pallavi Gusain†, Arundhati Jadhav‡,
Purabi Panigrahy§, Benjamin Stewart¶, Alekhya Penmatsa‖
and Tugrul Daim****

Portland State University, Portland, USA
*shivani2@pdx.edu
†pallavi.gusain@pdx.edu
‡ajadhav@pdx.edu
§panig2@pdx.edu
¶bstewar2@pdx.edu
‖alekhya2@pdx.edu
**tugrul.u.daim@pdx.edu

Most of the cities in US started making an effort to develop into smart cities. This chapter will discuss the current and future development of the technologies that provide solutions for traffic in different smart cities and help in the creation of a technology landscape analysis which will help Portland, Oregon, to formulate a combination of technologies and goals to achieve smart city goals. For the technology landscaping analysis, a thorough review of literature and an extensive study of use cases along with expert survey are done. This chapter aims to provide a highly detailed customized report and analysis of traffic sensors, trends, and recommendations for the

Smart City PDX Project. We provided a complete overview of many traffic sensor technologies to provide a strategic direction to traffic sensor technology investments for Smart City PDX. Our extensive expertise spans the traffic sensor technology life cycle from ideation to implementation. Based on the analysis of different case studies and along with the expert survey, it was found that using the latest technologies such as video imaging and thermal imaging will help Portland in preventing crashes and using data analytics will help in better traffic management.

1. Introduction

Our study is intended to create a Technology Landscape Analysis that will provide details on the current state of traffic solution technologies, trends in Smart City traffic solutions, and an idea of what solutions are on the horizon. This will be accomplished by looking at multiple US cities as case studies, identifying and comparing solution technologies to get a sense of their advantages and disadvantages, and collecting expert surveys on trends within smart city traffic solutions, the importance of different traffic solution technologies, and what helps a city implement their Smart City initiatives. Lastly, different scenarios tied to Portland city goals and pain points are outlined and plotted along with different technologies to see how they perform with respect to different scenarios.

2. Goals & Objectives

Portland city's primary goal is to make Portland's streets safe for everyone. The Traffic Safety Sensor Project will help PBOT achieve Vision Zero by 2025. The goal of Portland city is to completely get rid of road accidents and injuries that are caused by traffic (Safer Outer Stark, n.d.). Achieving Vision Zero requires rapid, effective action, which is only possible with accurate data about how people are using Portland's infrastructure, in particular people walking and biking, who are at highest risk of death or injury in crashes (Safer Outer Stark, n.d.). This chapter aims to achieve the following objectives:

(1) To learn different traffic sensor technologies.
(2) To study myriad parking sensor technologies.
(3) To build the use cases for Portland, Oregon, which impact the city and the risks associated with that scenario.

(4) To compare different the traffic sensor technologies in regard to advantages/disadvantages and applications.
(5) Analysis of experts' responses and to know future trends about traffic sensor technologies.

3. Literature Review

3.1. *Innovative data acquisition*

Measuring and converting the real-world phenomena into the digital signals which can be manipulated by software is known as "data acquisition". The Data Acquisition System (DAS) consists mainly of sensors, antennas, and other devices that can be used to record the action of the vehicle itself and its driver. The data that were recorded include the following:

(1) It describes the driver's and vehicle's actions.
(2) To make decisions, key data and safety application alert flags are used.
(3) This helps to interchange the information between the commercial vehicles and other nearby equipped vehicles.
(4) The basic safety application alerts are video image streams and cabin audio surroundings.
(5) The recorded data can monitor the vehicle functionality and health, so that any issues can be quickly detected and addressed.

The data acquisition systems are important for making transportation-related decisions because they are accurate and innovative systems. There are lots of problems that can be raised for both the policy makers and transportation practitioners due to a lack of accurate data.

3.2. *Review of machine learning and machine reasoning*

Recently, there have been numerous smart transportation applications that have been developed based on the rapid growth of the machine and deep-learning techniques. Big data is being generated in modern lives and in the current business world. This big data helps to improve the transportation system performance and at the same time helps to learn the user's behavior. The advancement of machine learning and machine reasoning has given us a chance to use the power of big data. By using machine-leaning techniques, the complex interconnections among the roadways, transportation traffic, environmental components, and vehicle crashes have been explored.

This chapter focuses on the major challenges faced by the transportation system, and by using machine learning techniques, these can be solved. It has been a decade since all the cities have started using innovative technologies such as Internet of Things (IoT), networking, and cloud computing to improve the long-term benefits to all the cities in the form of energy conservation (Aman Kumar, 2020).

Automation, machine learning, and IoT are the emerging technologies which are helpful in driving the smart city concept. There are various classic examples regarding smart cities such as parking and traffic management using smart technologies. Smart parking helps the drivers to know about the available parking through application without just roaming around for many hours in search of parking areas in populated city blocks. The smart meter also allows digital payment, so there is no chance of getting short of coins for the meter (Rouse, n.d.). Also, smart traffic management is used to manage the flow of traffic to reduce congestion during rush hour and it is helpful for public safety to reduce the number of accidents. Various sensor technologies are implemented to monitor the traffic in the current cities.

3.3. *Intelligent parking tools and techniques*

Intelligent parking tools consist of technologies, sensors, and applications which can be used to enhance parking efficiency and control the traffic, whereas intelligent sensor technologies are the tools which make it easy for a driver to find the vacant parking spaces. Some of the parking technologies that are mentioned in this chapter are active/passive infrared sensor, machine vision, inductive loop detectors, and magnetometers.

This chapter covers new and smart applications of machine learning techniques, such as wireless sensor networks, parking guidance systems, VANET, inductive proximity sensors, active ultra-sonic sensors, and radio frequency identification, to improve the challenges that are faced by the transportation system in the cities.

4. Traffic Solutions

One area targeted by smart city projects is improvements in traffic, including public transportation, private vehicles, delivery vehicles, bicycles, and pedestrians. In the future, this will include autonomous vehicles of various forms and uses. In 2020, it was found that the smart city projects significantly reduced the level of traffic congestion typically found in an urban environment (Guo *et al.*, 2020). The improvements were found to be the result of both the use of data to manage traffic problems and reductions from urban innovations implemented resulting from the smart city projects. An acceleration of improvements was acknowledged as governments continuously invest in smart city projects, noting that it takes time for results to be realized.

The US Department of Transportation started a smart city challenge asking mid-sized cities across America to expand ideas for an integrated, first-of-its-kind intelligent transportation system that would help people and goods to move speedily, cheaply, and efficiently by making use of data, applications, and technology (Smart City Challenge, 2017). Around 78 cities engaged in this competition. Besides spreading innovation, this competition aims to address issues of meeting the needs of residents of all ages and abilities, and bridging everything a city has to offer (Smart City Challenge, 2017).

Despite this, implementing technology requires careful planning and weighing of all factors, since there are many technologies currently available on the market, each with its own advantages and disadvantages. Petrick Beyer raised this concern in his research paper (Beyer, 2015). In this chapter, we are also going to touch upon the advantages and disadvantages of technologies with respect to Portland city's goals. This will provide Portland a holistic view of technologies that will benefit the Portland community best. As there are many technologies available in the market, all of which work in a similar way, the need for classification arises to simplify the use and have a better understanding of each technology. According to a study by Juan Guerrero-Ibáñez, Sherali Zeadally, and Juan Contreras-Castillo, traffic sensor technologies can be divided into two main types of technology: Intrusive/Pavement Invasive and Non- Intrusive/Non-Pavement Invasive (Guerrero-Ibáñez, 2018). The US Department of Transportation specifies that these traffic detectors can be used individually or in conjunction to measure variables such as presence, volume, speed, and occupancy.

While understanding which technology can be used to meet the goals and objectives, there is an equally important factor that is accuracy of these sensors at different places such as intersections and highways. As per the study conducted by Petrick Beyer, there have been many studies done on highway technology accuracy but very few for intersections. In

Table 1. Detection of accuracy for highways.

Detection technology	Accuracy (%)
Induction loop	99.3
Pneumatic road tube	99.4
Piezoelectric	99.9
Video	95.6
Magnetic	96.9
Radar	97.8
Infrared	95.5

Table 2. Detection of accuracy for intersections.

Detection technology	Accuracy (%)
Induction loop	97.9
Magnetic	96.4
Video	95.5
Radar	96.0
Infrared	92.0

his study, he found that intrusive technologies are better for short-term projects, while non-intrusive technologies work best for long-term projects of at least 3 years. As Portland's goals are generally long term, the recommendation would be to focus on non-intrusive technologies (Beyer, 2015). Tables 1 and 2 indicate that accuracy is better when using intrusive technologies, but overall, it is comparable to non-intrusive technologies considering other factors such as cost, installation time, traffic disruption, and life span (Beyer, 2015).

It could also be observed from Tables 1 and 2 that not all the technologies work the same for intersections and highways. So, our observation is that technologies are also usability dependent.

5. Traffic Solution Technology Backgrounds

5.1. *Passive infrared detectors*

Petrick Beyer in his paper, non-intrusive detection, mentioned that "A passive infrared detector does not emit any infrared light; it simply measures the level of infrared light emitted by vehicles. Infrared detectors can be used at the interjections to detect vehicles, to count vehicles, to measure speed and for classification" (Beyer, 2015).

Passive infrared detectors detect bicyclists and pedestrians based on heat signatures associated with human body temperature (Pedestrian and Bicycle Volume Data Collection Toolkit, n.d.).

5.2. *Active infrared detectors*

Infrared active detectors are like passive infrared detectors, except that they emit light in the infrared spectrum and then measure how much of the light is reflected to determine if there is a vehicle present in the field of view (Beyer, 2015). The sensor beam is sent between two devices, the sender and receiver (Pedestrian and Bicycle Volume Data Collection Toolkit, n.d.).

5.3. *Pneumatic tubes*

When a vehicle's tires pass over pneumatic road tube sensors, a burst of air pressure is sent along the rubber tube. By closing an air switch, the pressure pulse produces an electrical signal that is transmitted to a counter or analysis software. The pneumatic road tube sensor is portable, and powered by lead–acid gel or other rechargeable batteries (US Department of Transportation, 2014).

5.3.1. *Inductive Loop*

When a vehicle moves near or over a loop installed in the road surface, inductive loop detectors measure the change in its inductance. Inductive loops are used to detect vehicles at signalized intersections and to count vehicles (Beyer, 2015). Inductive loops are also used to collect bicycle data. Like traditional loop detectors used for signal detection and volume data collection, inductive loops are embedded into the road using diamond-shaped pavement cuts. To classify the count events, the sensors detect the presence of metal parts (Pedestrian and Bicycle Volume Data Collection Toolkit, n.d.).

5.4. *Piezoelectric*

According to prior research, piezoelectric sensors convert mechanical energy into electrical energy to collect data. A sensor is mounted in the road's surface and is used to count vehicles. Whenever a car drives over a piezoelectric sensor, it creates an electric potential — a voltage signal. The voltage signal is proportional to the degree of deformation. The voltage reverses when the car moves off. This can be used to detect and count cars. By the roadside, there is a counting device that is connected to the sensors. The data can be collected locally via an ethernet or transmitted via a modem (Retail Sensing, n.d.).

5.5. *Thermal imaging*

Thermography cameras record thermal radiation, which is produced by everything based on its temperature. As a result, you can determine the exact temperature of objects without even touching them. Thermo sensors are not affected by nighttime low light, excessive light from the sun, or shadows that can hide vehicles and pedestrians. Thermal sensors do not have any of these issues since they create a crisp image based on subtle differences in heat signature within a scene. Thermal sensors do not require any light to function, are not blinded by direct sunlight, and detect vehicles, pedestrians, and cyclists 24 hours a day, regardless of the amount of light available (Solutions for Traffic and Public Transportation Applications, n.d.).

5.5.1. *Video imaging*

Video image vehicle detection systems (VIVDS) are becoming increasingly popular at signalized intersections due to their convenient installation and rich information content. However, the current video imaging cameras must be installed at roadside poles or traffic light poles, which not only requires more than one camera to cover the entire intersection but also results in serious vehicle occlusions and adverse effects on vehicle detection and tracking (Wang *et al.*, 2009).

5.6. *Manual counting*

Manually counting vehicles is a simple but accurate method of traffic counting. People either use electronic handheld counters or tally sheets to record data. The manual method of counting vehicles is 99% accurate in tests (Vehicle Detection: Ten Ways to Count Traffic, 2019).

6. Parking Solutions

Parking is a universal pain point for drivers. People are frustrated and stressed during the search for a vacant parking area, with city drivers spending 30% of their time looking for spaces. A recent report by INRIX stated this by studying drivers' parking behaviors and experiences across 30 cities in the UK, US, and Germany (Smart Parking Solutions — IoT Sensors Space Race, n.d.).

Parking solutions provided by the government and Transportation department serve the purpose of preventing a wastage of time and fuel by working on Smart Parking solutions. These solutions also help in managing the traffic on busy roads by reducing any expected congestion due to slow-moving traffic.

Taking public transportation, such as buses and the metro, can help reduce parking-related congestion and air pollution. Nonetheless, private vehicles are still used for convenience. To reduce carbon emissions, we could replace all fuel-consuming vehicles with electric ones. Congestion would still be caused by the cruising of a vehicle to find a vacant parking space, even if we used electric vehicles. Perhaps driverless cars can alleviate this issue. Currently, driverless cars are in development. In certain conditions, companies such as Google have already developed prototypes that support a driverless mode. To replace existing vehicles, an automated driverless vehicle will require several technologies, including GPS, sensors, and V2V. Using smart parking sensors and technologies to detect parking occupancy information, which facilitates the drivers' decision-making process, is an alternative, not so far-fetched, solution. A variety of sensors and technologies can be used to gather parking occupancy information, including ultrasonic sensors, magnetometers, and multi-agent systems. Several smart parking applications are available online that provide directions to an empty parking space. These applications assist with improving parking efficiency.

In this chapter, we are going to investigate these smart parking IoT and AI-based solutions in detail, which are suitable for indoor parking spaces as well as outdoor parking spaces in cities.

6.1. *Smart parking tools and technologies*

6.1.1. *Wireless sensor network*

Through the integration of multiple devices, information can be gathered and processed from various sources to control the physical world. Sensor motes distributed in these networks work cooperatively to sense and control the environment (Paidi *et al.*, 2018).

6.1.2. *Parking guidance system*

Parking guidance systems are placed near the parking areas to make sure that the drives can clearly see the vacant spaces and then decide on which parking space to occupy. This system can also include a light on or off signal or a green or red color in different parking spaces. Light on or green means space is available, while light off or red means space is not available.

6.1.3. *VANET*

A Vehicular Ad Hoc Network provides services such as smart parking and antitheft with the help of wireless communication devices. In this network, Roadside units (RSUs) are placed across parking zones and vehicles where onboard units (OBUs) are installed. Here, a trusted authority is responsible for OBU and RSU registration. Therefore, once a vehicle approaches a parking zone installed with RSUs, the vacant parking space information will be provided to the OBUs.

6.1.4. *Inductive proximity sensor*

Unlike electronic proximity sensors, inductive proximity sensors do not require contact. Metal objects can be positioned and detected using inductive proximity sensors. The metal type determines the inductive switch's sensing range. The sensing range of ferrous metals, such as iron and steel, is usually longer, while the range of nonferrous metals, such as aluminum and copper, may be reduced by up to 60% (Proximity Sensor, 2014). A proximity switch is sometimes called an inductive sensor since it can output two possible states (Proximity Sensor, 2014).

6.1.5. *Active ultrasonic sensor*

Using reflected energy, these sensors would emit sound waves between 25 and 50 kHz. It can distinguish between a vehicle and a person based on the distance at which waves reflect. These sensors need to be placed on the top of every parking space to get the parking occupancy status.

6.1.6. *Radio frequency identification (RFID)*

Radio frequency tags are used to recognize vehicles. All the vehicles are given the radio frequency tag for recognition. A transceiver and an antenna are installed at the entrance of a parking lot to recognize the tag and permit the vehicle to occupy a parking zone.

6.1.7. *LIDAR*

This technology uses laser light reflections to identify the occupancy of a vehicle. This method has a great accuracy in the near range (up to ~7 m), but the accuracy greatly decreases with the distance increase and angle of arrival of the reflected signal.

6.1.8. *Camera detection*

To monitor the parking lot continuously, a camera sends out images at regular intervals and frame rates. Parking space occupancy can be detected by installing a camera overhead in a parking lot and using relevant image detection algorithms to segment vehicles and determine whether they are occupied.

6.1.9. *Magnetometer*

Detecting the presence of an object is done by measuring changes in electromagnetic fields. Therefore, magnetometers are placed beneath the surface so that they are close to the vehicle. There is no environmental impact. They can be used in both open and enclosed parking lots. It is possible to detect parking occupancy in real time using wireless sensors with a battery life of a few years.

7. Problem Statement

Traffic management is the focus area for all smart city dwellers and planners. Even though the focus of these smart cities is traffic management, traffic congestion, and avoidance, their approaches are different. Most of the cities in USA started developing as smart cities. We recognize that the programs that are good for one city/state may not necessarily be positive for another city, the environment, or the neighborhood. So, with multiple possible strategies and technologies, there is a need for standardized traffic solutions for smart cities in the USA. The decision-makers within each city must formulate a combination of technologies complying best with their smart city requirements and goals. There is also a lack of a summarization of the traffic solution technologies being used by other US smart cities for reference.

8. Methodology

The overarching methodology used will be the Technology Landscape Analysis. This method captures the current state of technology, best practices, and trends on data acquisition and data analysis. This will be accomplished by using multiple methods:

(1) *Case Studies*: Multiple US cities will be used as case studies. Key information will be the traffic solution vendor and technology, how they have used traffic sensors to work toward their smart city goals, and how successful their efforts have been.

(2) *Expert Surveys*: Surveys will be distributed to experts within academia, industry, and relevant organizations. The focus of the questions will be on future trends, success factors, and input on possible traffic solution technologies.

(3) *Technology Comparison Table*: Using the data gathered from various sources, each traffic solution technology will be identified along with its advantages, disadvantages, and application.

(4) *Use Cases*: Possible scenarios that could impact the city are explored along with what could trigger them. These cases are evaluated for what the risks or costs associated with leaving things as they are could be. Needed requirements for a traffic solution are identified for each case. Finally, a diagram maps out each use case along with the traffic solution technologies to help identify potential matches of case to technology. The seven cases that are discussed in this chapter will help explain the different scenarios that could cause the traffic congestion. The collected information can be used to analyze different machine learning and machine reasoning techniques that can be used to evaluate the congestion parameters.

9. Results

The US DOT has a list of smart cities. We picked some of the smart cities having the best traffic-related initiatives with proportional analysis of what type of traffic sensors they have and how they are using them. Refer to the Table A1 for the comparative analysis of traffic sensors used by different smart cities. The table mentions various sensors that are used by different states such as magnet, inductive loops, optical sensor, adaptive traffic signal control system, weather, cameras, transit priority, Bluetooth detectors, modularized wireless sensor arrays, camera and environmental sensors, advanced driver assistance system (ADAS), GPS technology, and secure FHSS.

We see that the data collected by these sensors are used for determining traffic characteristics, and we apply these details to the plan and performance of the traffic systems. The objectives of all these initiatives of different states include better data-driven decision-making, improved transportation, safer communities, and at the same time providing new economic development opportunities.

Given in the following are the insights into the traffic sensor initiatives of the cities we used for the comparative study.

9.1. *Portland, Oregon*

Portland has multiple smart city initiatives and goals (Smart City PDX, 2018) that it is pursuing under the guidance of Smart City PDX, a team which collaborates with the community, private and public sector partners, and the City of Portland staff.

One initiative directly relating to smart city traffic solutions is the Traffic Safety Sensor Project. It has completed a pilot program, partnering with Current by GE, AT&T, Intel, and Portland General Electric, to install 200 CityIQ nodes integrated into cobra head-style lighting. These sensors are installed on three of the most hazardous sections of road in the city. This pilot program is in support of "Vision Zero", Portland's goal to eliminate traffic accidents by 2025. Per the Portland Bureau of Transportation's website regarding the project, each of these CityIQ nodes can collect data on environmental, parking, and transportation activities. The sensors accomplish this via two cameras, an array of environmental sensors, CPU/GPU that allows for real-time analytics, a solid-state drive for storage, and cellular LTE for wireless data transmission.

Privacy has been at the forefront of Portland's considerations for smart city solutions. Following that guidance, each sensor analyzes images from the cameras to gather number, direction, and speed of pedestrians, bicyclists, and vehicles. This is done locally using the onboard sensors' computing capabilities. Only the data extracted from the images are sent to a data repository, Portland Urban Data Lake (PUDL), a storage place for the large amounts of raw data being created from various sensors and sources. The images from the cameras are not accessible and are deleted after data are extracted.

9.2. *Chicago, Illinois*

The Array of Things (AoT) was used by Chicago. It has now begun collecting a variety of real-time data on Chicago's environment and urban activities thanks to a network of sensor boxes mounted on light posts that were installed in 2016. In addition to installing a few sensors downtown in 2016, Chicago is now adding more sensors throughout the city, and the city's data portal currently shows the locations of all the AoTs installed and the yet-to-be-installed sensors (AoT, 2020). The city used modularized wireless sensor arrays. The standard components of the sensors are two cameras, a microphone, a computer, and a cooling solution. Additionally, 15 other modular sensors can be added. Temperature, barometric pressure, light, vibration, carbon monoxide, nitrogen dioxide, sulfur dioxide, ozone, ambient sound pressure, and pedestrian and vehicle traffic (speed, count, and direction) are all measured by modularized sensors. Engineers, scientists, policy makers, and residents will be able to use the data to work together to make Chicago and other cities healthier, more livable, and more efficient.

9.3. *New York, New York*

Connected Vehicle Application (CV) and Transit Signal Priority (TSP) are the smart traffic safety programs of New York City. Transit agencies can improve safety and traffic flow by installing smart devices across cities; their solution provides the power and connectivity needed to make it all happen (Global Traffic Technologies, 2018). Having smart devices installed in transit agencies will become increasingly important as transit agencies aim to improve service, create efficiencies, and improve quality of life for growing cities, and we are excited to be involved in this project. It primarily focuses on safety applications — which rely on vehicle-to-vehicle (V2V), vehicle-to-infrastructure (V2I), and infrastructure-to-pedestrian (IVP) communications (Briodagh, 2019). The applications alert drivers in advance to a possible crash to allow them to take steps to prevent the crash or reduce the severity of injuries and damage to vehicles and infrastructure.

9.4. *San Francisco, California*

Magnets and optical sensors are used by the city. Located on existing poles, real-time speed sensors (Miller, 2015) fill in the coverage gaps of loop detectors on California highways. We measure traffic speeds over a million times a day on this network, which operates 24 hours a day. Our server takes measurements several times per minute via the Cingular Wireless(R) data network with no latency, which are then formatted and delivered to data partners via the Internet. The city envisions shared mobility, i.e., the use of vehicles, public transportation, bicycles, and other modes of transportation jointly, that enables travelers' short-term access to transportation modes as needed. As part of its downtown parking capacity monitoring plan, the city is establishing a sensor network. Furthermore, they aim to create Vision Zero corridors that utilize vehicle technology for improved safety, health, and mobility.

9.5. Pittsburg, Pennsylvania

The city has the adaptable traffic signal control system, Split Cycle Offset Optimization Technique (SCOOT), Sydney Coordinated Adaptive Traffic System (SCATS), Real-Time Hierarchical Optimized Distributed Effective System (RHODES), and Optimized Policies for Adaptive Control (OPAC) "Virtual Fixed Cycle" and ACS Lite type of traffic sensors. Optimizing traffic flow is made possible by the sensors. Traffic origins and destinations between signalized intersections will be identified by this traffic detection system. It also helps in pedestrian detection and air quality monitoring capability (collecting real-time local ambient air levels of CO, CO_2, NO_2, SO_2, O_3, and PM (2.5) to better understand geospatial air quality "hot spots"). The city aims to become a leader in transportation innovation to address Pittsburgh's challenges and make all residents' lives better (SMART PGH, n.d.). The main purpose of the SmartPGH project is to expand opportunities for people, improve health by monitoring air quality, increase mobility, and promote safety for drivers and pedestrians.

9.6. Austin, Texas

The city consists of the following sensors: conventional traffic sensor technology, weather, cameras, transit priority, inductive loops, Bluetooth detectors, and boarding sensors. As Austin's Smart City vision takes shape, the sensors will increase safety by improving V2V communication, pedestrian/bicycle detection, intelligent sensor deployment to better manage incidents, smart station services that offer travelers a variety of safe options, and other components. As well as enhancing mobility, it creates ladders of opportunity, and it addresses climate change. The Texas capital builds on its history of innovative and successful public–private collaboration, in the open and flexible regulatory environment, our ongoing commitment to complete, compact, and connected communities, and the diversity and creativity of its citizens (Corridor News, 2020). Transportation agencies will have access to more and better data through the network of intelligent sensors and will be able to address a variety of problems facing local operations managers, public safety agencies, and planners.

9.7. *Columbus, Ohio*

The city consists of the Connected Columbus Transportation Network (CCTN) which includes traffic signals with traffic detection, DSRC, and pedestrian detection, truck loading zones with machine vision detection, parking and parking availability information, Wi-Fi hotspots, and transit service information (Bereton, 2019). The proposed approach is revolutionary in its unprecedented data integration, deployment of autonomous and connected vehicles, and integrating advanced sensors and cameras through smart intersections. The project will create economic development and job opportunities, as well as ladders to greater access to jobs, fresh food, services, education, healthcare, and recreation for residents. ParkMobile is an event parking management app that is adopted by the Columbus to free up the traffic caused by vehicles that are waiting for parking spaces.

10. Technology Comparison Tables

10.1. *Traffic solution technology comparison table: Pedestrians and bikes*

Table A2 summarizes research focused on technologies capable of measuring pedestrian and bicycle traffic, including sensors such as active infrared detectors, passive infrared detectors, inductive loop, pneumatic tubes, piezoelectric, thermal imaging, video imaging, and manual counting. Bicyclists and pedestrians are detected when the beam transmission is broken using active infrared, with the difference being the sensor beam is sent between two devices (sender and receiver). Both the sensors cannot work in the presence of debris or snow, and they detect pedestrians and bikes as mixed traffic because they cannot distinguish between them (Pedestrian and Bicycle Volume Data Collection Toolkit, n.d.). Pneumatic tubes can detect only shared-use and bicycle lanes. It can also distinguish between motorcycles and bikes. However, they are sensitive to snow and get damaged very easily while road sweeping or snow plowing. Inductive loops are also a bicycle-specific data collection technology. Diamond-shaped pavement cuts are used by setting the Inductive loop into the gravel. To count the events, the sensors are arranged to detect the presence of metal parts of a bicycle. Most of the inductive loops can work in shared-use paths and on-street mixed-traffic situations. Careful planning is required to set up inductive loops as they are not reusable once fitted under the street. Piezoelectric detectors are an exception among all the detectors mentioned above in that they can detect the speed and direction along with counting of the vehicles. It has a high complexity of installation(Pedestrian and Bicycle Volume Data Collection Toolkit, n.d.).

Overall, most of these technologies are not appropriate for detection of both pedestrians and bicyclists. Therefore, options for Portland city seem very limited, among which three stand out: thermal imaging, video imaging, and manual counting. Manual counting is not of much help as one needs to manually count the number of pedestrians and bikers crossing. This results in high chances of human error as well as not being able to help in emergencies, making predictions, etc. However, Portland city did use manual counting to validate the sensors. Overall, thermal imaging and video imaging seem to be the most viable solutions. Thermal imaging has advantages over video imaging in two main aspects — first, it helps with privacy protection for people and, second, it is effective in any weather conditions as well as light conditions. These two aspects make it more fitting for achieving Portland city's goals.

10.2. Traffic sensor technology landscape analysis: Vehicles

In Table A3, we classified traffic sensors into two types, intrusive and non-intrusive (*Traffic Control Systems Handbook*: Chapter 6. Detectors, 2017). It could be concluded based on above analysis that in intrusive sensors, traffic flows are disrupted, while on the other hand for non-intrusive sensors no traffic disruption is required (Juan Guerrero-Ibáñez, 2018). Alas, there are some advantages of using intrusive sensors over non-intrusive and the most important advantage is technology maturity (Juan Guerrero-Ibáñez, 2018). Also, they have been in existence for long periods which means they have been widely implemented already in many places and have high accuracy in detecting vehicles. However, according to our analysis, in the long run, disadvantages far outweigh the advantages. Prolonged use of intrusive sensors leads to high maintenance costs, traffic disruption, as well as high installation cost, in comparison to non-intrusive sensors (Juan Guerrero-Ibáñez, 2018).

Among non-intrusive sensors, microwave radar detector, ultrasonic detector, and acoustic detector have limitations to their efficacy in some situations. For example, the microwave radar is unfit for the large steel structures. Similarly, both ultrasonic and acoustic detectors do not work well on freeways (Beyer, 2015). On the other hand, video cameras and thermal imaging work well everywhere else and are less susceptible to temperature changes. When video cameras are used with nearby IR, it becomes a passive sensor which is much more beneficial than any other sensor in Portland's scenario (Meritee, 2010).

10.3. *Traffic sensor technology landscape analysis: Connected vehicles*

From Table A4, we can observe that connected vehicles are categorized into three categories: V2I (Vehicle to Infrastructure), V2V (Vehicle to Vehicle), and V2X (Vehicle to Everything) (Gettman, 2020). Table A4 also gives insights into advantages/disadvantages and application of each of the Connected Vehicle Technologies. Based on Table A4, it is evident that V2I works better for short to middle ranges and it has a slower data rate. So, it will not be a better fit for Portland city (What is Vehicle to Infrastructure V2I Technology? n.d.). However, V2V is meant to work for large ranges and is more efficient in traffic management, prevention of crashes, reducing congestion, and optimizing traffic routes as well (What is Vehicle to Infrastructure V2I Technology? n.d.). However, it is prone to hacking and fatal consequences due to failure.

V2X seems to be the solution for all the problems as it connects vehicles to community as well as to infrastructures. But, it is also prone to the same disadvantages as V2V and privacy is a major concern here which needs to be addressed (Gettman, 2020). Nevertheless, it is an emerging technology and is least expensive of all; therefore, Portland should consider this for implementation once it is out of the prototyping phase.

10.4. *Parking sensor technology landscape analysis: Vehicles*

We have categorized parking sensors and technologies into two types, outdoor and indoor parking lots, based on the suitability of working in covered parking lots and open parking lots. This categorization has helped us decide which technology Portland should go for to improve its Smart parking system. We summarized the information in Table A5.

Our analysis to select the parking technology from Table A5 is based on the information from different papers (Smart Parking Solutions — IoT Sensors Space Race, n.d.; Jeffrey Joseph, 2014; Idris, 2009).

Camera Detection technology can be most effective for both outdoor and indoor parking lots. This includes continuous frames captured by image processing to detect the vehicle on the spot by differentiating between subsequent frames. This can deliver accurate detection of the vehicles only when the view of the camera is within the detection zones and when there is better visibility due to light. Various external and environmental factors like inclement weather, shadows, vehicle projection into adjacent spots, and day-to-night transition can affect the performance of this technology. These limitations can be overcome by using virtual 3D-scene information and Machine Vision (Rendering an Image of a 3D Scene: An Overview, 2009–2016).

As Portland City has both indoor and outdoor parking lots, this technology has proven to be the most cost-effective one out of all given in Table A5. RFID, Active Ultrasonic, and LIDAR sensors' functions can be affected by external environmental factors and, among these, LIDAR sensors are the most expensive. Wireless Sensor Networks are low-cost technology, but being wireless these are vulnerable to security attacks and sometimes can deliver low communication speed over the network which can result in the incorrect information update. VANETs need dedicated tamper-resistant communication devices in each vehicle and some older vehicles do not have these so they cannot participate in these networks. Under the parking areas, the inductive proximity sensors need to be installed, which deteriorates the life of the pavement. On the same note, Magnetometers are costly to install and maintain on a huge scale. The Parking Guidance System can only be beneficial for indoor parking lots when the vehicle is already in the parking.

Therefore, camera detection can be the technology system that is easiest to maintain and manage for indoor and outdoor parking lots. Thus, it can participate as a traffic management solution in managing the traffic and keeping the parking spot seeking vehicles off the road.

11. Expert Panel

Experts were identified and asked to complete a survey to gain knowledge regarding general trends within smart city traffic solutions, details about areas that should be focused on when cities are implementing smart city initiatives, and feedback on specific traffic solution technologies.

To ensure that the results of the survey represent experts of varied backgrounds and disciplines, experts from industry, academia, government, and non-government organizations were contacted for their input. Table A6 outlines the composition of the expert panel used. As shown, the panel represents a balanced mix of the four targeted expert categories. Appendix A lists the questions the expert panel was asked to answer (Figure A1).

The experts' ratings for the various technologies were given numeric values depending on their answer. "Most Important" was given four points, "Very Important" three points, "Important" two points, "Slightly Important" one point, and "Not Important" zero points. Answers of "I'm Not Sure" were omitted in the aggregate scores for each technology. The top five technologies identified as the most important for smart cities were video imaging, V2I, V2X, V2V, and thermal imaging.

The experts were also asked to identify the major trends within smart city traffic solutions that they would expect to see in the next 5–10 years. These trends are important to take into consideration as cities are identifying their traffic solutions to ensure they take full advantage of anticipated future changes. The top four major trends, which were found in at least 30% of the expert surveys, are connected vehicles (V2V, V2I, and V2X), active traffic management/ATSPMs, improvements to machine learning and data analytics, and autonomous vehicles.

12. Use Cases

Use cases represented in Table A7 identify possible scenarios where smart city traffic sensors may play a role. The possible risks associated with maintaining current systems and abilities are identified along with how traffic sensors could play a role in minimizing the impact of these cases.

The data need to resolve each use case as well as the potential smart city traffic solution technologies onto the same chart. The horizontal axis identifies the types of measurement data that can be gathered by each sensor, including the type of traffic measured (pedestrian, bicycle, vehicle, or all three) and if the sensor identifies count only or speed and direction as well. The vertical axis identifies two pieces of information: for how long the technology gathers the data as well as if the installation and service are intrusive, requiring a roadway to be temporarily closed, or non-intrusive. In general, technologies closer to the upper-right corner will satisfy more scenarios. Technologies marked with an asterisk require two instances of the technology to be used in tandem to gain the direction and speed data in addition to the number of vehicles or bikes.

Use Cases 1, 5, and 6 are satisfied by video imaging or thermal imaging. Use Cases 2, 3, and 4 do not require data on bikes or pedestrians, allowing them to be satisfied by microwave radar, ultrasonic, acoustic, inductive loop, piezoelectric, or magnetometer traffic solutions. Case 7 is satisfied by manual counting. Ultimately, video imaging and thermal imaging can satisfy all seven of our identified use cases.

13. Considerations When Designing a Solution

From our survey responses, our experts shared certain pitfalls regarding the traffic sensors. They are listed in the following along with our suggested solutions to deal with these issues:

(1) *Data Management*: By storing all structured and unstructured data in a centralized repository, the data lake will be able to handle any size of dataset. By storing data as is, without the need to structure it first, and running various types of analytics, such as dashboards and visualizations to big data processing, better decisions are guided.

(2) *Cooperation Between Stakeholders*: There are so many people involved in the project and they want to be informed from the very beginning through to application and appraisal and they all need to be communicated with differently. A proper communication plan must be created in conjunction with the project team. There should be a focus on the communication needs of the stakeholders by using a suitable technology to authorize more productive and systematic communications with stakeholders and others.

(3) *Data Privacy and Security*: To protect data and act in accordance with data protection laws, there is a need for both Data Privacy and Data Security. Privacy can become a way to engage with people and show them that their data are respected. Utmost care should be taken to store and dispose of personal information securely. Data encryption can be done wherever and whenever possible.

(4) *Community Involvement and Buy-In*: In fact, the reality is that no matter how much time is spent on minute details of a project, there will always be elements that could be missed during the planning process. Most of the evidence-based interventions come from community engagement and it helps in better decision-making.

(5) *Appropriate Selection of Technology Complying with the Requirement of the Smart City Project*: Our technology landscaping analysis will help in solving this issue to a greater extent.

14. Recommendations

Based on our analysis and experts' evaluations, we have come up with the following recommendations for Portland smart city:

(1) *Using the latest technology* such as video imaging and thermal imaging will help Portland in preventing crashes and in better traffic management using data analytics.

(2) *Connected vehicle technology* is going to be the most prominent trend soon. So, our recommendation is for Portland city to start building up a plan to accommodate C-V2X technology. This tech will help in bringing the communities closer to work with the city's goal and vision.

(3) *Open data for public* will enhance the chances of getting better and least expensive solutions from the public. However, privacy factors need to be taken care of before giving the public the access to the data. This will help in creating a competitive environment and thus getting the best results.

(4) *Technology maturity* evaluation is very important, especially for Portland city. Portland faced the problem of not checking the technology maturity time and ended up taking a trial version of the tech which was ultimately not as beneficial as they expected. Therefore, collaborating with other Smart Cities that have already implemented some of the solutions or researching the case studies or projects already executed by the vendors becomes essential. Working with renowned and leading vendors like FLIR should be the primary approach.

(5) *System engineering analysis* is also advised to get a clear understanding of technology along with implementation technicalities to keep the existing techs and new technology in sync and to perform efficiently overall.

15. Peek into the Future

With the advancements of new technologies, smart roads will be reshaped. New technology developments and applications, such as the application of data analytics, deep learning, and artificial intelligence, are emerging in smart transport.

In the future, roads will no longer be seen as stable infrastructures, but rather a smart grid which will be fully aware of the surrounding conditions, framework, and the environment. Smart roads will be totally combined end to end, which is an advantage for city residents, leaders, and operators. People are allowed to interconnect with the smart transport system through their smartphones. The city operators are free to monitor traffic and environmental circumstances remotely and carefully by answering on time on a demand basis. Eventually, smart cities will progress toward the goal of achieving near-zero mortality and hence modernizing our lives for the better.

16. Suggestions for Future Research

We recognized certain limitations to our chapter, and we believe the following points will be key ideas for future research:

(1) Inclusion of solutions provided by vendors in technology landscaping analysis.
(2) Inclusion of initial cost, installation cost, and maintenance cost analysis in the decision-making process.
(3) Reassessing the model recommended from our technology landscaping analysis.
(4) More research into the solution for connected vehicles.
(5) To enhance the performance of the current traffic system in the cities, it is better to use the cloud-based frameworks.
(6) It is better to make use of IoT for traffic management. Moreover, systems such as automated vehicle systems, advance driving active safety (ADAS), and real-time network optimization are some of the machine learning techniques that address the challenges faced by transportation.

17. Conclusion

The goal of any city is to use its resources most effectively with smart city initiatives and projects when moving forward. Many potential traffic solutions have been identified throughout this chapter, and evaluated for their advantages and disadvantages, rated with respect to importance by experts and correlated with multiple use cases focused on Portland's goals and areas of concern. Taking this information into consideration as well as the variety of traffic solution combinations being explored by the top smart cities in the US, multiple solutions could be implemented to meet city goals.

However, all technologies are not equally important for a smart city traffic solution. Consistently throughout our chapter, in comparison tables, expert surveys, and use case analyses, the two technologies that can gather the most complete data, and therefore satisfy the most potential goals, have been video imaging and thermal imaging. Other technologies could be used to supplement video imaging and thermal imaging or used when focusing on goals with lesser data requirements. By clearly identifying city goals and the corresponding data that need to be captured to meet those goals, the selection of the best traffic solution technology can be simplified and narrowed down dramatically using our Traffic Solution Comparison Tables, Expert Opinions, and Use Case Graph.

References

Aman Kumar, J. S. (2020 May 13). A journey from conventional cities to smart cities. Intechopen: https://www.intechopen.com/books/smart-cities-and-construction-technologies/a-journey-from-conventional-cities-to-smart-cities.

Array of Things. (2020). https://arrayofthings.github.io/.

Bereton, E. (2019 Dec 11). State Tech. https://statetechmagazine.com/article/2019/12/street-smarts-how-columbus-advancing-its-smart-city-infrastructure.

Beyer, P. (2015). Non-intrusive detection, the way forward. Repository: https://repository.up.ac.za/bitstream/handle/2263/57785/Beyer_Intrusive_2015.pdf?sequence=1&isAllowed=y#:~:text=Advantages%20of%20a%20Pneumatic%20road,disruption%20to%20traffic%20during%20installation.

Briodagh, K. (2019 Nov 26). Smart city IoT project launches for New York city DoT. Smart Transport Feature News: https://www.iotevolutionworld.com/smart-transport/articles/443661-smart-city-iot-project-launches-new-york-city.htm.

Chai, K. & Toh, J. A. (2020 Jan 22). Proceedings of a royal society. https://royalsocietypublishing.org/doi/10.1098/rspa.2019.0439#d455375e1635s.

Citron, R., Samms, G., & Woods, E. (2019). Smart city tracker 2Q19. Navigant Research: https://guidehouseinsights.com/-/media/project/navigant-research/navigant-research-executive-summaries/tr-scit-2q19-executive-summary.pdf.

Corridor News. (2020 Jun 26). https://smcorridornews.com/austin-to-pilot-smart-city-technology-to-improve-congestion/.

DHM Research. (n.d.). Portland transportation priorities — Survey report. The city of Portland: https://www.portlandoregon.gov/transportation/article/480189.

Gettman, D. (2020 Jun 3). DSRC and C-V2X: Similarities, differences, and the future of connected vehicles. Kimley-Horn: https://www.kimley-horn.com/dsrc-cv2x-comparison-future-connected-vehicles/.

Global Traffic Technologies. (2018 Jan). https://www.gtt.com/project/new-york-mta-signal-priority/.

Guerrero-Ibáñez, J., Zeadally, S., & Contreras-Castillo, J. (2018 Apr 16). Sensor technologies for intelligent transportation systems. NCBI: https://www.ncbi.nlm.nih.gov/pmc/articles/PMC5948625/.

Guo, Y., Tang, Z., & Guo, J. (2020). Could a smart city ameliorate urban traffic congestion? A quasi-natural experiment based on a smart city pilot program in China. *Sustainability*, 12(6), 2291–2310.

Idris, M. Y. I., Leng, Y. Y., Tamil, E. M., Noor, N. M., & Razak, Z. (2009). Car park system: A review of smart parking system and its technology. *Information Technology Journal*, 8(2), 101–113. Science Alert.

Jeffrey Joseph, R. G. (2014). Wireless sensor network based smart parking system. *Sensors & Transducers*, 162(1), 5–10.

Juan Guerrero-Ibáñez, S. Z.-C. (2018 Apr 16). Sensor technologies for intelligent transportation systems. PMC: https://www.ncbi.nlm.nih.gov/pmc/articles/PMC5948625/.

Meritee, A. (2010 Aug 23). How to use passive infrared sensor cameras. ACTi Knowledge Base: https://www2.acti.com/getfile/KnowledgeBase_UploadFile/How_to_Use_Passive_Infrared_Sensor_Cameras_20101015_004.pdf.

Miller, B. (2015 Oct 28). California cities turn to Internet of Things to solve parking, traffic problems. FutureStructure: https://www.govtech.com/fs/California-Cities-Turn-to-Internet-of-Things-to-Solve-Parking-Traffic-Problems.html.

Paidi, V., Fleyeh, H., Håkansson, J., & Nyberg, R. G. (2018 Apr). Smart parking sensors, technologies and applications for open parking lots: Review. *IET Intelligent Transport Systems*, 12(8), 735–741.

PBOT Portland Bureau of Transportation. (n.d.). The City of Portland: https://www.portlandoregon.gov/transportation/40390.

Pedestrian and Bicycle Volume Data Collection Toolkit. (n.d.). CODOT gov: https://www.codot.gov/programs/bikeped/documents/nmm_toolkit.pdf.

Proximity Sensor. (2014 Dec 7). Slideshare.net: https://www.slideshare.net/arif0o7/proximity-sensor-42438958.

Rendering an Image of a 3D Scene: An Overview. (2009–2016). Scratchapixel 2.0: https://www.scratchapixel.com/lessons/3d-basic-rendering/rendering-3d-scene-overview/rendering-3d-scene.

Retail Sensing. (n.d.). People Counting Systems: https://www.retailsensing.com/definition/piezoelectric-sensor.html.

Rouse, M. (n.d.). Smart city. IoT Agenda: https://internetofthingsagenda.techtarget.com/definition/smart-city#:~:text=A%20smart%20city%20is%20a,government%20services%20and%20citizen%20welfare.

Safer Outer Stark. (n.d.). PBOT: https://www.portlandoregon.gov/transportation/78832.

Smart City Challenge. (2017 Jun 29). U.S. Department of Transportation: https://www.transportation.gov/smartcity.

Smart City PDX. (2018 Jun 19). https://www.smartcitypdx.com/traffic-safety-sensor-project.

Smart Parking Solutions — IoT Sensors Space Race. (n.d.). Newark.com: https://www.newark.com/smart-parking-solutions-the-iot-sensors-space-race#.

SMART PGH. (n.d.). http://smartpittsburgh.org/.

Solutions for Traffic and Public Transportation Applications. (n.d.). Intelligent Transportation Solutions: https://www.flirmedia.com/MMC/CVS/Traffic/16-1639_US.pdf.

Traffic Control Systems Handbook: Chapter 6. Detectors. (2017 Feb 1). Office of Operations: https://ops.fhwa.dot.gov/publications/fhwahop06006/chapter_6.htm.

U.S. Department of Transportation. (2014 Nov 7). Policy and Governmental Affairs: https://www.fhwa.dot.gov/policyinformation/pubs/vdstits2007/04.cfm.

Vehicle Detection: Ten Ways to Count Traffic. (2019 Jun 4). Retail Sensing: https://www.retailsensing.com/people-counting/vehicle-detection-count-traffic/#:~:text=Manual%20counts&text=A%20person%20either%20uses%20an,accurate%20over%20the%20counting%20period.

Wang, Y., Zou, Y., Shi, H., & Zhao, H. (2009 Sep 22). Video image vehicle detection system for signaled traffic intersection. IEEE Xplore: https://ieeexplore.ieee.org/document/5254300.

What is Vehicle to Infrastructure V2I Technology? (n.d.). rgbsi: https://blog.rgbsi.com/what-is-v2i-technology.

Appendix A: Expert Survey Questions

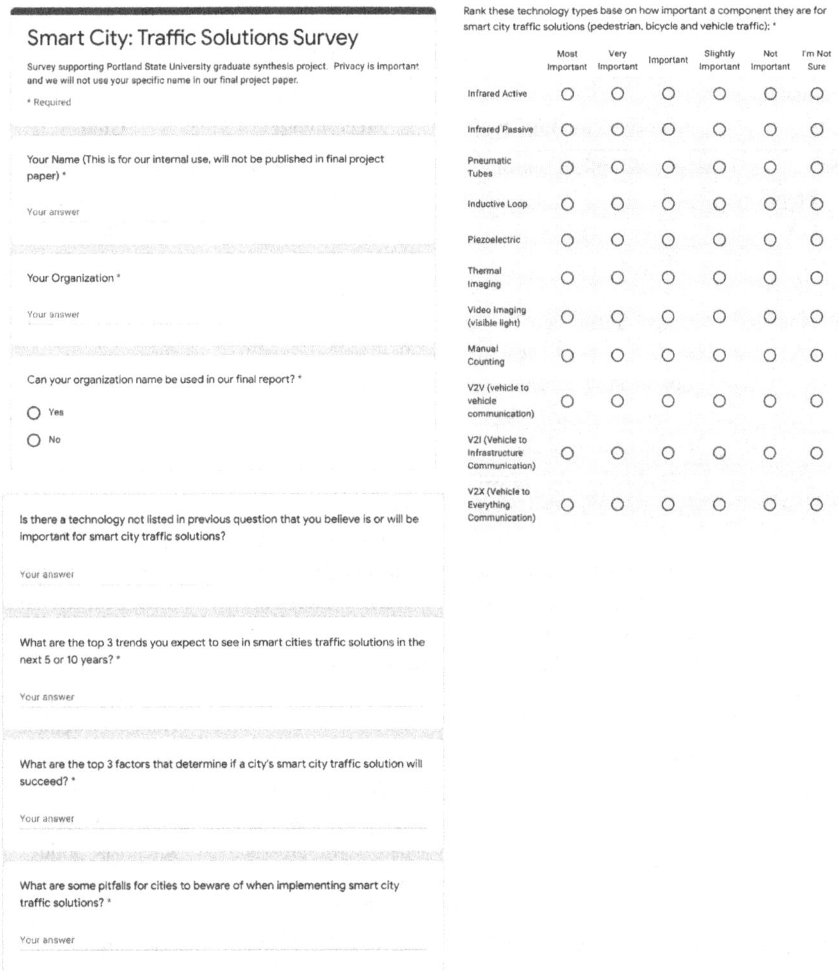

Smart City: Traffic Solutions Survey

Survey supporting Portland State University graduate synthesis project. Privacy is important and we will not use your specific name in our final project paper.

* Required

Your Name (This is for our internal use, will not be published in final project paper) *

Your answer

Your Organization *

Your answer

Can your organization name be used in our final report? *

- Yes
- No

Is there a technology not listed in previous question that you believe is or will be important for smart city traffic solutions?

Your answer

What are the top 3 trends you expect to see in smart cities traffic solutions in the next 5 or 10 years? *

Your answer

What are the top 3 factors that determine if a city's smart city traffic solution will succeed? *

Your answer

What are some pitfalls for cities to beware of when implementing smart city traffic solutions? *

Your answer

Rank these technology types base on how important a component they are for smart city traffic solutions (pedestrian, bicycle and vehicle traffic): *

	Most Important	Very Important	Important	Slightly Important	Not Important	I'm Not Sure
Infrared Active	○	○	○	○	○	○
Infrared Passive	○	○	○	○	○	○
Pneumatic Tubes	○	○	○	○	○	○
Inductive Loop	○	○	○	○	○	○
Piezoelectric	○	○	○	○	○	○
Thermal Imaging	○	○	○	○	○	○
Video Imaging (visible light)	○	○	○	○	○	○
Manual Counting	○	○	○	○	○	○
V2V (vehicle to vehicle communication)	○	○	○	○	○	○
V2I (Vehicle to Infrastructure Communication)	○	○	○	○	○	○
V2X (Vehicle to Everything Communication)	○	○	○	○	○	○

Figure A1. Survey questions.

Appendix B: Tables

Table A1.   Smart cities' summary.

Name of city	Sensor/technology type	Use of sensor/ technology	Goals
San Francisco, CA	Magnet & optical sensor.	The speed measurements are sent to a server via the cingular wireless(R) data network.	Shared mobility, sensor network, and connected vision zero corridors.
Pittsburgh, PA	Adaptive traffic signal control system — Split cycle offset optimization technique and optimized policies for adaptive control and integrated control systems.	Optimizing traffic flow and air quality monitoring.	Expands opportunities for people and promote safety.
Austin, Texas	Conventional traffic sensor technology, sensors for loops, weather, cameras, transit priority, Bluetooth detectors, boarding sensors.	Connected vehicle technology, vehicle/ bicycle detection, enhance mobility.	Builds innovative and successful public safety and intelligent sensors.
Columbus, Ohio	Truck loading zones with machine detection of zones and multifunctional kiosks.	Advanced smart sensor technology.	Data integration, autonomous, and connected vehicle deployment.
Chicago, IL	Modularized wireless sensor arrays.	Modularized sensors measure temperature, barometric pressure, light, vibration, and gases.	To provide data to help engineers, scientists, policy makers, and residents.
New York City	Advanced driver assistance system, GPS technology, secure FHSS.	Making vehicle travel time faster.	Reduces traffic congestion and improve intersection operation.
Portland, OR	Camera and environmental sensors.	Identifying vehicle, pedestrian, and bicycle traffic.	Pilot new sensor technology deployed to streetlights to gather the data.

Table A2.   Traffic solution technology comparison table.

Technology sensor	Advantages	Disadvantages	Application	Vendors
Infrared Active	– Currently classifies vehicles into eight categories. Capable of detecting and classifying pedestrians and bicycles.	– Does not automatically distinguish between pedestrian and bikes. Needs a separate algorithm for pedestrian detection. – Sensitive to ambient background temperatures.	– Used to know the direction of travel and count of pedestrians passing. – Sidewalk or shared-use path.	– TRAFx – EcoCounter – TrailMaster
Infrared Passive	– Relatively low cost and out-of-the-box capability.	– Does not automatically distinguish between pedestrian and bikes. – Less effective than active infrared.	– Used to know the direction of travel and count of pedestrians passing. – Sidewalk or shared-use path.	– TRAFx – EcoCounter – TrailMaster
Pneumatic Tubes	– Portable and relatively easy to deploy.	– Due care should be taken to avoid damage from vandalism or routine maintenance, such as street sweeping and snow plowing. – Not appropriate for use during snow season.	– Ideal for short duration count of bikes passing over the tube on the road.	– EcoCounter – MetroCount – TRAFx – Road Sys

(Continued)

Table A2.   (*Continued*)

Technology sensor	Advantages	Disadvantages	Application	Vendors
Inductive Loop	– Can work in both shared-use paths and on-street mixed-traffic situations.	– The loops use magnetic fields for detection, they are sensitive to utility lines, either overhead or buried, so careful planning is needed. – Not reusable.	– On-road or paved shared-use path.	– Eco-Counter – Road Sys
Piezoelectric	– Helpful where bikes cannot travel in bike-dedicated lanes.	– The complexity of installation includes higher precision cuts and the installation of a utility box to house the processing and data storage equipment.	– The devices are capable of measuring bicycle volume, direction, and travel speeds.	– Metro-Count
Thermal Imaging	– Simple efficient and discrete surveillance. – Effective during the day and at night. – High performance in all weather and air conditions (rain, fog, snow, smoke). – Low maintenance for lower cost of ownership. – Motion and human activity detection. – Protection of privacy. – Fewer false alarms.	– The higher sticker prices. – Thermal images cannot be captured through some materials like water and glass. – Undesirable individuals are detected but not identified.	– Provides uninterrupted 24-hour detection of pedestrians and bicyclists regardless of the amount of light available.	– FLIR

Video Imaging	– Can be used for car, pedestrian as well as bike identification. – Neural network-based technology. – Capable of detecting and classifying bicycles and pedestrians. – Real-time analysis.	– High cost/hour of data collection, but optimal attribution. – Lightning and weather conditions can affect the video imaging. – Maybe restriction based on privacy concerns.	– Processes video data and classifies moving objects.	– Miovision Ubicquia/ CityIQ – Array of things
Manual Counting	– High accuracy. – Can count pedestrians, bicycles, and vehicles.	– Very costly and time-consuming. – Data limited to narrow time windows.	– Accurately monitors pedestrians, vehicles, and bicycle traffic for a short period of time.	– N/A (City Employees)

Table A3.   Traffic sensor technology landscape analysis: vehicles.

Sensors type	Technology	Advantage	Disadvantage	Application
Intrusive	Inductive loop detector	– Low cost and power consumption. – High accuracy.	– Poor detection of small vehicles. – Damage by road deterioration or heavy vehicles. – Major disruption to traffic during installation. – Sensitivity to temp fluctuations.	Vehicle detection and counting.
	Pneumatic road tube	– Low cost. – Quick to install. – Low power consumption.	– Poor detection of small vehicles. – Damaged by road deterioration or heavy vehicles. – Major disruption to traffic during installation. – Sensitivity to temperature fluctuation. – Affected by metallic road construction material and high risk of the loop and feeder cable theft.	Used for keeping track of the number of vehicles, vehicle classification, and vehicle count.

	Type	Advantages	Disadvantages	Use
	Magnetometers	– Less susceptible than loops to stresses of traffic. – Insensitive to inclement weather such as snow, rain, and fog. – Some models transmit data over wireless RF link.	– Installation requires pavement cut. – Improper installation decreases pavement life. – Installation and maintenance require a lane closure. – Models with small detection zones require multiple units for full lane detection.	Useful on bridge decks and viaducts, where the steel support structure interferes with loop detectors, and loop can weaken the existing structure. Magnetometers are also useful for temporary installations in construction zones.
	Piezoelectric	– Low power consumption and high accuracy.	– Detection can be affected by temperature changes. – Damaged by road deterioration. – Major disruption to traffic during installation.	Classification of vehicles, vehicle count, and measuring vehicle's weight and speed.
Non-intrusive	Video cameras	– Quick to install. – Very reliable. – Video monitoring. – Flexible setup. – Optional uni-directional operation. – No disruption to traffic during installation. – Real-time analysis.	– High initial cost and detection accuracy affected by occlusion.	Detection of vehicles across several lanes and can classify vehicles by their rate, occupancy, and speed for each class.

(Continued)

Table A3.   (Continued)

Sensors type	Technology	Advantage	Disadvantage	Application
	Thermal imaging	– Less expensive in long run due to less downtime, power outages, production losses, fires, etc. – Very accurate. – Can see in any light conditions, day or night. – See through smoke/dust/fog, etc. – Does not produce images for others to see, so no privacy concerns.	– Image requires some training to interpret. – Long start-up time. – Expensive initially. – Larger/heavier.	Helps you clearly see road hazards in total darkness, and it will alert drivers to nearby vehicles, people, infrastructure to take necessary steps.
	Microwave radar detector	– Sensitive to reflect microwave from the objects, without interference from temperature. – Low power consumption. – Constant/pulse wave operation mode.	– CW doppler sensors may not detect stopped vehicles. – Some models have problems when used in large steel structures (i.e., steel bridges). – Overhead conductors within the beam cone can cause problems.	Automatically adjusts the time of traffic conditions in order to optimize traffic flow.

Ultrasonic detector	– Multiple-lane operation available, capable of over height vehicle detection. – Large Japanese experience base.	– Environmental conditions such as temperature change and extreme air turbulence can affect performance. Temperature compensation is built into some models. – Large pulse repetition periods may degrade occupancy measurement on freeways with vehicles traveling at moderate to high speeds.	Vehicle speed detectors where the speed of the vehicle is used as a reference in providing a decision about the condition of the road.
Acoustic detector	– Passive detection. – Insensitive to precipitation.	– Large pulse repetition periods may degrade occupancy measurement on freeways with vehicles traveling at moderate to high speeds.	Measuring vehicle passage, presence, and speed by passively detecting acoustic energy or audible sounds produced by vehicular traffic.

Table A4.   Traffic sensor technology landscape analysis: connected vehicles.

Sensors type	Technology	Advantage	Disadvantage	Application
Connected vehicles Tech	V2I	– 80% of road accidents could be avoided. – improves safety, convenience, and efficiency.	– Data rate is slower than Wi-Fi or Bluetooth. – Short to middle range only.	Sharing of vehicle data with infrastructure owner-operators (IOO) (cities, counties, and states responsible for traffic operations).
	V2V(DSRC)	– Fully designed. – Verified to be viable. – Tested in major pilots in New York, Florida, and Wyoming. – Improves fuel efficiency via truck platooning. – Optimizes routes. – Prevents crashes. – Improve traffic management and reduces congestion.	– DSRC could not cover as much area as C-V2X. – More expensive than C-V2X. – Privacy of the owners and users of the vehicles is a major concern. – Prone to hacking. – Fatal consequences during failure of the system. – Liability concerns. – Expensive.	Transmission of data is vehicle-to-vehicle.
	V2X(C-V2X)	– Less expensive to manufacture. – Provides an upgrade path for faster and higher-performance electronics.	– Still in the prototype phase. – Privacy of the owner and users of the vehicles is a major concern. – Prone to hacking. – Fatal consequences during failure of the system.	Used for very high-speed and high-frequency data exchange (up to 10 times per second with millisecond latency).

Table A5. Parking sensor technology landscape analysis: vehicles.

Technology	Advantage	Disadvantage	Application	Suitability for outdoor/indoor* parking lots
Wireless sensor network	Flexibility, inherent intelligence, low cost, rapid deployment, and more sensing points.	Vulnerable to security attacks, nodes need to be charged at regular intervals, low communication speed.	Real-time data collection from a parking space for users to reserve a parking spot in advance.	Outdoor and indoor
VANET	A system no longer needs a sensor at each parking spot as it can use strategically placed transmitters to triangulate the location of the vehicles to determine if it is occupying a parking spot.	Requires the vehicles to have a dedicated tamper-resistant communication device. This means older vehicles would not be compatible with the system without modification.	Prevention of collisions, safety, blind crossing, dynamic route scheduling, real-time traffic condition monitoring, etc. Another important application for VANETs is providing internet connectivity to vehicular nodes.	Outdoor and indoor
Inductive proximity sensor	Very good at detecting metal objects over the sensor (such as vehicles) and will not send a false positive if someone walks over the sensor.	It needs to be buried under the parking spot and would require the parking spot to be unavailable during installation, also resulting in a decrement in pavement life.	Used for detecting vehicles entering or leaving a parking lot.	Outdoor and indoor

(Continued)

Table A5.   (*Continued*)

Technology	Advantage	Disadvantage	Application	Suitability for outdoor/indoor* parking lots
Magnetometer	Insensitive to inclement weather such as snow, rain, and fog.	Expensive to install and maintain on a large scale.	Detects the presence of a vehicle by detecting the change in an electromagnetic field.	Outdoor and indoor
Cameras detection	Accurate detection by image processing (if detection zones are within the field of view of the camera and good lighting is available), ease of management and maintenance, cost-effective.	Privacy issues, needs analytics for image processing, inclement weather, shadows, vehicle projection into adjacent spots, day-to-night transactions can affect the performance.	Continuous frames captured by the video image processor can be used in the detection of vehicles as it reveals the difference between subsequent frames.	Outdoor and indoor
Parking guidance system	Saves time and gas.	Only beneficial for people already in the parking and not for people on the way to destinations.	Finding the free parking with the light off for occupied.	Outdoor
RFID sensor	Would not detect other objects, such as animals or people.	It requires the RFID to be present on the vehicles.	Used to detect if a vehicle is occupying a parking spot.	Indoor

| Active ultrasonic sensors | Very good at detecting objects in the parking spot. | Unable to distinguish between vehicles and other objects. | Used to calculate the distance to the object based on the frequency of the pulse. | Indoor |
| LIDAR sensor | The sensor is more accurate and works at a farther distance than RFID. | Much more expensive than RFID. | Determines the distance to the object. | Indoor |

Notes: *Indoor: Totally covered. Not much impact of weather and external environmental factors.
*Outdoor: Open parking lots. Not covered. Impacted by external environmental factors.

Table A6.   Expert panel.

Expert #	Organization	Category
Expert 1	Ubicquia	Industry
Expert 2	Urban. Systems Inc.	Industry
Expert 3	Boulder AI	Industry
Expert 4	Portland State University	Academic
Expert 5	Portland State University	Academic
Expert 6	Transportation Research and Education Center (TREC) at Portland State University	Academic
Expert 7	Technology Association of Oregon (TAO)	Non-Government Organization
Expert 8	Open Commons	Non-Government Organization
Expert 9	Portland Bureau of Transportation	Government Organization
Expert 10	Portland Bureau of Transportation	Government Organization

Table A7. Uses cases.

Case name	Case triggering event	Current response	Opportunity cost or risk	Data from traffic solution
Case 1: Increase in traffic-related death and serious injury (PBOT Portland Bureau of Transportation, n.d.)	Portland currently monitors and tracks traffic-related deaths and injuries as part of their Vision Zero goal. Vision Zero aims to eliminate traffic deaths and serious injuries by 2025.	Limited data on traffic characteristics (pedestrian, bike, vehicle) from manual counting, pneumatic tubes, video imaging on 3 streets.	Without widespread data to analyze, the root cause or improvements that could reduce traffic fatalities and major injuries may not be apparent to city decision-makers.	Long-term data to see historically how traffic has changed would allow comparison to change in traffic deaths and injuries. Pedestrian, bicycle, and vehicle traffic data on speed, quantity, and direction.
Case 2: Increased traffic congestion (Jaquiss, 2020)	I-5 between Portland and Vancouver is typically congested by 2 or 3p.m. weekdays, slowing to 20 or 25 mph. Many other areas experience similar congestion from events, changing traffic flows, and road closures.	Tolls are a potential change that could be implemented. A recent lack of I-5 congestion due to Covid-19 indicates a toll may be capable of reducing congestion by managing demand.	The current system of aging sensors is inaccurate and cannot reliably gather data on volume of traffic using I-5 between Portland and Vancouver.	Historic and current data on vehicle quantity, speed, and direction. Accurate data can identify the efficacy of policy changes, such as potential tolls or an additional bridge.

(Continued)

Table A7. (*Continued*)

Case name	Case triggering event	Current response	Opportunity cost or risk	Data from traffic solution
Case 3: Public transportation delays (DHM Research, n.d.)	Portland TriMet consists of bus, light rail, and commuter rail. Delays can occur from traffic congestion, inclement weather, machine failure, or blockages of required pathways.	Increasing number of buses and Max trains, dedicated lanes for bus traffic.	Many Portland citizens rely on public transportation. Delays can lead to lower user confidence, causing ridership declines and an increase to vehicle traffic.	Historic and current data on vehicle quantity, speed, and direction. This data can be compared with data on public transportation delays.
Case 4: Impact of new traffic-related change	When the city makes a change to policy, signage, laws, traffic signals, etc., they need to know the impact of those decisions to determine if the changes had the desired results.	Limited data from pneumatic tubes, manual counting, and the 3 roads with video sensors. The impact is difficult to ascertain.	Without sufficient data, efficacy of changes cannot be known. This is especially important for pilot programs that could be implemented city-wide.	Historic and current data on vehicle quantity, speed, and direction. Any traffic-related changes can be evaluated to see if the expected results are achieved.
Case 5: Allocation of traffic-related infrastructure funding	Portland would like to use funding for traffic-related infrastructure improvements to have the greatest impact on Portland's citizens.	Portland can use deaths/injuries, citizen input, etc., to determine where to make changes.	Without actual data, areas of increased risk to pedestrians, bicyclists, or vehicles could be overlooked.	Current data to see how infrastructure is being used. Pedestrian, bicycle, and vehicle traffic data on speed, quantity, and direction.

Case 6: Informing community	The Portland community travels using many modes of transportation. Having access to information on the current and historic state of streets, sidewalks, etc., would allow them to best navigate the city safely.	Limited data on traffic volume, speed, and direction are available to citizens to best navigate Portland safely.	Bicyclists, pedestrians, or vehicles may travel streets or sidewalks with higher risks or more congestion, causing higher likelihood of injury or delays.	Historic and current data on vehicle, pedestrian, and bicycle quantity, speed, and direction would allow app developers to fully utilize the data gathered.
Case 7: Verifying new sensors	Any new sensor technology for measuring some aspect of traffic must be confirmed to be sufficiently accurate.	Manual counting is currently used to determine accuracy.	Manual counting is very short in duration. It can verify accuracy within a tight window, allowing for possible inaccuracies during differing conditions (high/low traffic, rain, snow, fog, high/low temps, etc.).	Manual counting is required currently; however, a long-term method of verification would be preferred.

Part 3

Data Engineering

Chapter 10

Mitigating the Proclivity Toward Multiple Adjustments Through Innovative Forecasting Support Systems

Banusha Aruchunarasa* and **H. Niles Perera†**

Center for Supply Chain, Operations and Logistics Optimization, University of Moratuwa, Katubedda, Sri Lanka
*banushaa@uom.lk
†hniles@uom.lk

Extant literature scarcely addresses the use of multiple adjustments in supply chain forecasting. In order to come up with innovative insights to address this, this chapter looks into the magnitude of awareness, the reasons behind the multiple forecast adjustments, and the elements that contribute to it. The authors deploy an international survey to understand why Forecasting Support Systems are subjected to multiple adjustments. The responses underscore the fact that the extant Forecasting Support Systems have failed to win the confidence of most forecasters. This study finds the reasons driving multiple adjustments and the individual impact of each reason toward multiple adjustments. Promotions are found to be the prime motivator for multiple adjustments, while government regulations and competitive actions too are major considerations. Further, the findings accentuate that there is a sizable belief that multiple adjustments lead to better forecast accuracy. Incorporating innovative features into

Forecasting Support Systems is vital to mitigating multiple adjustments and improving forecast accuracy. These findings provide fertile soil to be considered when developing innovative Forecasting Support Systems that can reduce the propensity for multiple adjustments.

1. Introduction

Forecasting is the lifeblood of the supply chain (Bruzda, 2019). Judgmental forecasting is widely observed in the industry (Connor, 1992). This relates to humans contributing to the generation of forecasts (Perera *et al.*, 2019). Empirical research states that judgmental adjustments are a common practice in forecasting where the forecaster adjusts a system-generated forecast (Lawrence *et al.*, 2006). The forecasting process has several stages starting from forecasting data to the fine-tuned final stage. In every stage, the human does several adjustments due to various contextual factors which are defined as information, except time series and expert knowledge that can be used to extrapolate, interpret, and anticipate the behavior of the time series (Önkal *et al.*, 2008). In many cases, the choice of selecting an appropriate forecasting method itself is judgmental (Perera *et al.*, 2019).

This manuscript considers an underreported forecasting issue. Multiple adjustments in forecasting are common and obviously refer to modifying a forecast more than once by multiple stakeholders (Önkal *et al.*, 2008). This is not well addressed in academia and industry (Perera *et al.*, 2019). Therefore, the key research problem is to investigate the magnitude of multiple adjustments to forecasts in the industry and the factors driving this. In this research, various factors were investigated in terms of the perception of forecasting professionals involved in multiple industries scattered all over the world. This research was conducted to gain a clear understanding of the magnitude of multiple adjustments in the industry. Moreover, this chapter hopes to understand why forecasters do not trust forecasts generated by Forecasting Support Systems (FSS) with the expectation of developing more informed FSS with higher efficacy.

2. Literature Review

The literature review starts with two major recent systematic reviews in the research content relevant to this study (Arvan *et al.*, 2019; Perera *et al.*, 2019). Thereafter, the literature review was driven by a keyword search initiated on major research repositories such as Scopus and Google

Scholar. The search includes keywords such as forecasting support systems, multiple adjustments, judgmental adjustments, behavioral supply chain management, and judgmental forecasting. Thirty relevant papers are reviewed in this chapter coming from top journals including *Decision Sciences*, *International Journal of Forecasting*, *European Journal of Operational Research*, and *Production and Operations Management* covering the fields of manufacturing, operations research, and supply chain management. The authors found that extant literature scarcely addresses the use of multiple adjustments in supply chain forecasting and only two papers addressed the research area regarding multiple adjustments (Önkal *et al.*, 2008; Perera *et al.*, 2019).

2.1. *Forecasting*

Forecasting is the process of predicting the future by using past and present data (Mehrjerdi, 2016). It is a socio-technical system that includes humans, digital devices, and established processes (Phillips & Nikolopoulos, 2019). Forecasts are a vital part of the supply chain. Most of the supply chain functions are triggered by forecasts (Van den Broeke *et al.*, 2018). From a business point of view, forecasts play an important role as they are used to derive budgeting using a market plan (Klassen & Flores, 2001). In the industry, there are job roles under names such as demand planner, sales forecaster, and analyst. Despite the semantic differences in their role, they provide forecasts that drive other operations of the supply chain (Boone *et al.*, 2019).

Boone *et al.* (2019) listed several issues as the main challenges to the development of accurate forecasts. The presence of disaggregated forecasts, the selection of appropriate supply chain data, data collection points from all the nodes in the supply chain, the influence of new data sources from the upstream and downstream, the uncertainty factors related to the supply chain partners, the impact of forecast errors, and the effects of linking forecasting to supply chain decisions are among them. There are several publications providing solutions to the abovementioned forecasting issues faced by supply chain partners (Arvan *et al.*, 2019).

2.2. *The human factor in supply chain forecasting*

Based on empirical studies, it is proven that forecasting relies on human judgment (Sanders *et al.*, 2003). Forecasters depend on hyper-rational actors based on self-interest, consciousness, and objective-oriented behavior (Bendoly *et al.*, 2010). The process and the relevant human resources are closely connected at each level in supply chain forecasting (Tangpong *et al.*, 2019). Boudreau *et al.* (2003) analyzed some industry cases and stated the importance of integration of technical and human considerations into the business processes as well as FSSs and provided a behavioral operational framework based on human factors such as capability, opportunity, motivation, and understanding.

Perera *et al.* (2019) provide a systematic literature review and bibliometric analysis of human factors in supply chain forecasting. Based on the bibliometric analysis, they report six distinct clusters to structure the research agenda in judgmental forecasting. They are conventional forecasting methods, the impact of retail promotions, forecasting support systems, impact on supply chain operations, industry-focused research, and behavioral constructs in forecasting. In each cluster, they provide areas where there are knowledge gaps that are important to the furtherance of the discipline. In their sixth cluster behavioral constructs in forecasting, they provide an analysis of forecasting based on behavioral aspects. Multiple adjustments are one such area discussed under this cluster. This is the core focus of this chapter.

2.3. *Forecast support system*

Support systems are designed to analyze and integrate the available information to support and incorporate human judgment (Perera *et al.*, 2019). FSSs are one type of decision support system (DSS) which helps to develop forecasts (Fildes *et al.*, 2006). The main features of an FSS are a time series database with event histories, quantitative forecasting techniques, and the facilities to integrate human judgment into the forecasts (Fildes *et al.*, 2006; Matharage *et al.*, 2020). Webby *et al.* (2005) discuss the forecasters' ability to incorporate contextual information in forecasting by utilizing judgmental adjustments through Forecast Support Systems (FSSs). There is solid evidence from recent studies that forecasts derived from forecast support systems are routinely and subsequently adjusted by the forecasters based on managerial judgments (Arvan *et al.*, 2019; Fildes & Goodwin, 2021). Further, these judgmental interventions are harmful to the forecast accuracy and consume a considerable amount of management time (Fildes & Goodwin, 2021). Judgmental interventions should only be incorporated into the FSS where the contextual information is present with the forecasters, such as information about forthcoming events, promotions, and competitor actions that is not captured in the statistical method (Fildes & Goodwin, 2021; Fildes *et al.*, 2018; Sanders, 1992). These adjustments can be individual adjustments or group adjustments (Zellner *et al.*, 2021).

Therefore, integrating the adjustments with FSS would help improve forecast accuracy and reduce the time wastage of forecasters (De Baets & Harvey, 2018). Defining an efficient forecast support system by incorporating judgments will create maximum benefits to the supply chain while increasing forecast accuracy and time utilization (Fildes & Goodwin, 2021).

The likelihood of implementing FSS to reduce the forecasters' time to do the forecast depends on the cognitive behavior of the forecaster (Goodwin *et al.*, 2018). One of the major research gaps identified by Goodwin *et al.* (2018) is to find ways to mitigate the distrust on FSS toward improving algorithm aversion. FSS should integrate the appropriate combination of the statistical methods and judgments without judgmental biases (Fildes *et al.*, 2006).

This chapter considers the nature of multiple adjustments and the reasons to integrate the multiple adjustment aspect into FSS.

2.4. *Judgmental adjustments*

Empirical research studies state that human adjustments are a common practice in forecasting. Judgmental adjustments are a commonly observed forecasting practice in order to improve forecast accuracy (Eroglu & Croxton, 2010). Sanders & Manrodt (1994) did a study of 240 firms in the US related to forecasting and accuracy. Their study reveals that 60% of firms that used forecasting software were using judgmental adjustments to correct the forecasts generated through the system routinely. Hence, understanding the appropriate use of judgmental adjustment is vital for a forecaster (Lawrence *et al.*, 2006).

Forecasters conduct adjustments considering several factors related to the forecasting environment. The adjustments are mostly related to and dependent on the forecaster's intuition, experience, and informational advantages. Adjustments are made to achieve the operational goals in several sectors as supply chain management, revenue management, logistics, and quality management (Önkal *et al.*, 2008). Judgmental adjustments are beneficial in the case of improving forecasting accuracy under two conditions. The first condition is whether the statistical forecast is insufficient to identify the time series pattern. The second condition is whether the forecaster has vital contextual information which cannot be captured by systematic models (Lawrence *et al.*, 2006).

Judgmental adjustment of a system-derived forecast occurs in two steps. The first step is determining the requirement of adjustments to the statistical forecast and the second step is determining the direction and size of the adjustment (Lawrence *et al.*, 2006). The minimization of asymmetric loss in the forecasting process is considered as one of the main reasons behind judgmental adjustments in organizations (Goodwin, 1996), where the asymmetric loss in supply chains relates to the asymmetry of the loss due to stocking out as opposed to overstocking since stocking out is considered more costlier than overstocking. These asymmetric losses make forecasting systems sensitive to the real impacts of forecast errors (Lee, 2007). Despite these benefits, judgmental adjustments can be inefficient in the case of heuristics and biases and empirical research studies report that inefficient judgmental adjustments lead to big losses in forecasting. Therefore, the organization can benefit from human judgment to achieve their business targets (Perera *et al.*, 2019).

2.5. *Multiple adjustments in forecasting*

Industry observations and feedback reveal that multiple adjustments are practiced frequently. It is a key research gap highlighted in the recent systematic review on the human factor in supply chain forecasting (Perera *et al.*, 2019). In many industries, there are common situations where the forecasters receive inputs that have already been adjusted by another forecaster and the forecasters often adjust the received forecasts using judgmental adjustments. Understanding multiple adjustments is vital for the organization as it can improve forecast accuracy. This includes adjustments repeated for the same reasons by different forecasters (Önkal *et al.*, 2008).

Önkal *et al.* (2008) conducted two studies to identify two main pillars of adjustments related to this area. The objectives of these studies were to identify the possible result of adjustment framing and identify the ripple effects of the provision of explanations behind adjustment with already adjusted forecasts. Önkal *et al.* (2008) reported the finding of their first study as evidence of previously adjusted forecasts being adjusted less than the original forecasts. Additionally, the finding of their second study supports providing reasons alongside the previously adjusted forecasts to inform further adjustments. The practitioners are willing to accept previously adjusted forecasts than unadjusted forecasts to be used as an input in their operations (Lawrence *et al.*, 2006; Önkal *et al.*, 2008). The accuracy of adjustments of previously adjusted forecast depends on changes in temporal information and the understanding of the forecaster about the forecasting task. Environmental factors such as communication, collaboration, and transparency of information also impact multiple adjustments (Perera *et al.*, 2019).

Trust is one of the main drivers of multiple adjustments. The performance of multiple adjustments depends on trust, bargaining power, and level of information sharing. Extant literature states that trust is the main reason behind the lack of effectiveness of forecast adjustments. Trust is important in situations like uncertainty and asymmetric information (Ghosh & Fedorowicz, 2008). Information sharing and the quality of information sharing depend on the trustworthiness between forecasters. Trust plays a key role in multiple adjustments in forecasting (Ghosh & Fedorowicz, 2008).

Extant literature recommends a three-step mechanism for multiple adjustments in forecasting which is identified as a key research gap (Perera *et al.*, 2019). The three steps are to identify the reason behind forecasting, the modifications rationale, and the analysis of previous adjustments. Empirical work states that adjusted forecasts are more accurate than unadjusted forecasts (Önkal *et al.*, 2008). The method of presenting the forecasting data also creates an impact on the adjustments. The visual presentation of the previously adjusted forecast data impacts the next forecasters' decision on forecast adjustments (Perera *et al.*, 2019). Only a few studies have addressed the adjustments of previously adjusted forecasts and further works have been encouraged to fill this void (Önkal *et al.*, 2008; Perera *et al.*, 2019).

3. Methodology

Surveys are deployed since there is a gap in understanding the factors leading to multiple adjustments. This also helps to plug the vacuum of research studies on multiple adjustments in recent years obstructing a proper understanding of the topic. The authors identified the need for establishing whether it is a serious phenomenon worthy of academic investigation before more rigorous methods are employed. Literature (Dalrymple, 1987; Fildes & Goodwin, 2007; Sanders & Manrodt, 1994; Sanders *et al.*, 2003) use surveys to understand the practice of judgmental forecasting. The research survey was developed with inspiration from these studies.

The questionnaire consists of 21 questions related to multiple adjustments in forecasting. Each question in the questionnaire focuses on different aspects of multiple adjustments in forecasting. There are four sections to the questionnaire. The four sections cover the research problem and all the objectives of the research.

Section 1 — General information.
Section 2 — Information related to forecasting.
Section 3 — Information related to judgmental adjustments.
Section 4 — Information related to multiple adjustments.

Through the literature, it is identified that multiple adjustments to forecasts are not well recognized and addressed in the industry. Forecasting professionals of an organization are those responsible for the adjustments

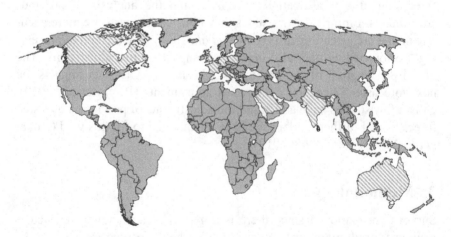

Figure 1.   Distribution of respondents based on country of affiliation.

made to forecasts. The survey targets forecasting professionals from different companies across the globe. Once the survey design is finalized, the survey was propagated among forecasting professionals who are executive grade or above in supply chain organizations or academics with forecasting experience. The survey was posted publicly in the social media accounts of the International Institute for forecasting to collect responses from around the world. All valid responses are considered as the research sample. 135 responses were received from participants from 16 countries representing 4 continents. Figure 1 presents the distribution of the responses from different countries. The authors understand that there were regional biases in the collected data as a major part of the responses are from Sri Lanka.

4. Data Analysis and Findings

The responses include Manufacturing, Transport, Logistics, Maritime, Education, Consulting, Retail, Apparel, Electrical, Media, Arts and Crafts, Construction, and Government sectors. The collected data were analyzed using descriptive statistics, Principal Component Analysis (PCA), and Cronbach's Alpha test through the SPSS analytical software.

The awareness of using multiple adjustments in the forecasting industry is not exclusively addressed in the literature. Therefore, the research problem is to investigate the magnitude of practicing multiple adjustments to forecasts and the factors leading to multiple adjustments.

Starting from the general findings of the analysis related to forecasting, the forecasting process is mostly (78%) done by executive-level employees in the industry. Other managerial positions do not interact with the forecasting process as much in an operational context as per the findings. There is also evidence of multiple adjustments being more prolifically used in manufacturing industries such as fast-moving consumer goods and apparels as opposed to other industries. This is due to the continuous flow of demand information downstream. Most of the industries use system-generated forecasts in their operations, but despite the technological improvements, there are some companies still using manual forecasts in their operations. Microsoft Excel is playing a major role in computing forecasts as a system/software tool. Demand solutions, R Studio, and SAP are some of the popular systems used by the industries presently.

A majority of the respondents (76%) stated that they are aware of the forecasting model their employer uses. This finding can be considered as an antithesis for the feasibility of the black box theory (Perera *et al.*, 2019). Exponential smoothing is a well-established forecasting method used by many FSSs in the industry to compute forecasts. Other established methods like ARIMA, Box — Jenkins, Holt Winter, and Naïve are also used in the industry as per the findings. The analysis shows 36% of the forecasters are preparing forecasts monthly and some forecasters forecast through a combination of daily, weekly, monthly, quarterly, and annual buckets, based on their requirements (32%). This agrees with past research findings (Fildes *et al.*, 2018). The analysis shows the Mean Absolute Percentage Error (MAPE) is the error measurement method that is widely used by forecasters to compute forecasting accuracy. The responses reveal that some companies are using more than one method to compute forecast accuracy for benefits including simplicity, ease of interpretation, and communication. Hence, the combination of different error measurement methods is useful when comparing forecast performance across different datasets (Fildes & Goodwin, 2007).

Judgmental adjustments are known by 78% of the respondents in the industry. Similarly, 72% of the respondents state they practice judgmental adjustments. The analysis shows forecasters who adjust system-generated forecasts practice judgmental adjustments more than 25% of the time on average. This provides ample evidence of the inadequacy of FSS to capture the trust of forecasting professionals.

A majority (69%) of the respondents stated that they are aware of multiple adjustments to system-generated forecasts. A large portion (65%) of the respondents agreed that they are practicing multiple adjustments to system-generated forecasts and 28% stated that they are not practicing multiple adjustments to system-generated forecasts. 7% of respondents stated that they are unaware of the practicality of multiple adjustments in their supply chain. The respondents who are aware of multiple adjustments have higher forecasting experience (more than 2 years) and possess a higher tier of employment in their respective organizations.

Figure 2 shows the results of this question in which the respondents were asked to choose how often they do multiple adjustments in their industrial operations. A majority (26%) of the respondents stated that they adjust forecasts less than 25% of the time for already adjusted forecasts. The second-highest number of respondents (23%) stated that they do multiple adjustments between 50% and 75% of the time. 21% of the respondents stated that they do multiple adjustments more than 75% of the time.

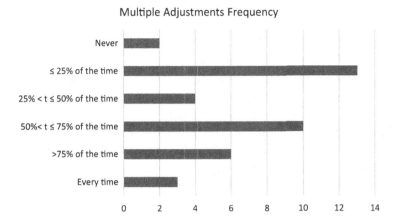

Figure 2.   Adjustment frequency.

The authors also observe from the survey results that a considerable number of industry forecasters (14%) do multiple adjustments to every forecast they touch. This indicates a serious lapse in FSS. This also supports past findings reported in the literature (Lawrence *et al.*, 2006).

There are 11 reasons for multiple adjustments that were identified through literature and industry responses (Lawrence *et al.*, 2006; Önkal *et al.*, 2008; Petropoulos *et al.*, 2016) which were listed down in the research questionnaire with a Likert scale of 1–7 where 1 = low impact and 7 = high impact. Principal Component Analysis (PCA) was employed to identify the principal component for multiple adjustments and rank the reasons based on the impact. All the reasons identified from the research are the perceptions of forecasting professionals toward the concept of multiple adjustments. The analysis provides a ranked output according to the impact of the reasons behind multiple adjustments. The reasons are mentioned in the following in order of decreasing importance:

(1) Promotional and advertising activities.
(2) Changes in regulations/government policies.
(3) Competitive actions.
(4) Effects of sudden occurrences.
(5) Price changes.
(6) Economic trends.
(7) Environmental factors.
(8) The previous person has not considered all the factors.
(9) Seasonal factors.

(10) Experience.
(11) The previous forecaster usually over-forecasts/under-forecasts.

Therefore, the highest contributing reason behind multiple adjustments is promotional and advertising activities. Also, other major contributions are due to changes in government regulations and competitive actions.

The respondents were asked to choose between yes and no to identify whether multiple adjustments lead to better forecast accuracy. Around 91% of the respondents stated that they believe multiple adjustments lead to better forecast accuracy. This finding suggests that there is a sizable belief in multiple adjustments leading to better forecast accuracy (Önkal et al., 2008). It also shows that there is less confidence in system-generated forecasts among practitioners as well as highlighting the importance of developing innovative FSS derived through the methodical analysis of extant forecasting adjustment practices.

Figure 3 shows the result when respondents were asked to choose between the options for the level of accuracy improvement due to subsequent (multiple) adjustments. A majority (37%) of the respondents stated that the level of accuracy improvement as a result of the subsequent adjustments is more than 75%. The authors further observe from the survey results that a considerable number of industry forecasters (3%) stated that the level of accuracy improvement as a result of the subsequent adjustments is 100% accurate. This could well be a vote of confidence on one's own ability. Finally, 5% of the respondents stated that there is no accuracy improvement because of subsequent adjustments. Additionally,

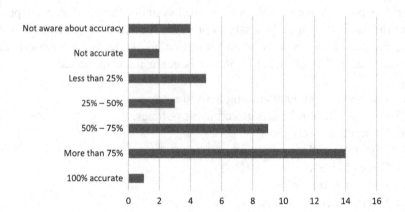

Figure 3.   Accuracy improvement.

10% of the respondents stated that they are not aware of the level of accuracy improvement because of the subsequent adjustments.

While the sample size was not large enough to compare the results in different industry sectors, the above findings suggest that the globally dispersed respondents tend to show trust toward subsequent adjustments leading to accuracy improvements as opposed to FSS developed through rigorous procedures. Future research should seek to discern whether this belief is true. If yes, it is important to understand why a higher level of accuracy could not be achieved through system-generated forecasts or a single adjustment. This highlights the importance of FSS developers paying more attention to addressing this critical gap and introducing FSSs which are practical and capable of reducing the need for multiple adjustments while assuring high levels of accuracy. It would also be beneficial to investigate whether group forecasting would yield more accurate outputs as opposed to allowing multiple adjustments by different individuals at different stages of the forecasting process.

5. Conclusion

This research provides a preliminary investigation of multiple adjustments in forecasting. This study determines the awareness and magnitude of multiple adjustments through a survey. Experienced forecasters in the industry stated that they practice multiple adjustments 25% of the time. There are 11 reasons identified as drivers of multiple adjustments. The highest contributing reason is identified to be promotional and advertising activities. Changes in government regulations and competitive actions are other key drivers.

The survey provides foundational insights about the presence of multiple adjustments to forecasts in the industry which can be used as a launch pad for improving the FSS and for future research. The awareness of multiple adjustments, tracing multiple adjustments in forecasting, and providing suitable guidance to the forecasters regarding multiple adjustments are some of the factors which can be seamlessly integrated into the FSS during the development process. The FSS can be designed to capture forecast adjustments and reasons for consideration during subsequent adjustments while also guiding forecasters. An innovative FSS can be developed by considering the trends aligned with multiple adjustments, reasons driving multiple adjustments, and the accuracy improvement from the multiple adjustments.

This study contributes to the literature as well as to the industry to identify the proclivity toward multiple adjustments in supply chain forecasting. The findings provide direction to the supply chain industry to improve forecasting accuracy and improve FSS to be more robust to effectively blend innovative technology with human expertise. Additionally, this work is applicable for any industry which involves inventory management and forecasting. Therefore, it can be applied to any business process where any promotional activities and other contextual factors are present. This research problem can arise in any module of an ERP system. Further work in this avenue, such as the impact of the black box effect in multiple adjustments, an algorithm to integrate multiple adjustments into FSS, the relationship between the behavioral aspects and multiple adjustments, and the effect of data visualization in forecast support systems, requires further investigation. Furthermore, an experimental study is welcomed to validate the research findings of this study. Moreover, future work can also investigate the effect of multiple adjustments on inventory decisions as the intersection of forecasting and inventory is vital to the success of a supply chain (Balachandra *et al.*, 2020; Perera *et al.*, 2020).

Acknowledgments

This research was funded by Grant ID SRC/LT/2020/20 by the Senate Research Committee of the University of Moratuwa, Sri Lanka. The authors are grateful to Prof. Dilek Önkal and Dr. Share De Beats for their generous support toward this research.

References

Arvan, M., Fahimnia, B., Reisi, M., & Siemsen, E. (2019). Integrating human judgement into quantitative forecasting methods: A review. *Omega (United Kingdom)*, 86, 237–252. https://doi.org/10.1016/j.omega.2018.07.012.

Balachandra, K., Perera, H. N., Thibbotuwawa, A. (2020). Human factor in forecasting and behavioral inventory decisions: A system dynamics perspective. In Freitag, M., Haasis, H. D., Kotzab, H., & Pannek, J. (eds.), *Dynamics in Logistics*. LDIC 2020. Lecture Notes in Logistics. Cham: Springer.

Bendoly, E., Croson, R., Goncalves, P., & Schultz, K. (2010). Bodies of knowledge for research in behavioral operations. *Production and Operations Management*, 19(4), 434–452. https://doi.org/10.1111/j.1937-5956.2009.01108.x.

Boone, T., Boylan, J. E., Fildes, R., Ganeshan, R., & Sanders, N. (2019). Perspectives on supply chain forecasting. *International Journal of Forecasting*, 35(1), 121–127. https://doi.org/10.1016/j.ijforecast.2018.11.002.

Boudreau, J., Hopp, W., Mcclain, J. O., & Thomas, L. J. (2003). Commissioned paper on the interface between operations and human resources management. *Manufacturing & Service Operations Management*, 5(3)(September 2015), 179–202. http://dx.doi.org/10.1287/msom.5.3.179.16032%0AFull.

Bruzda, J. (2019). Quantile smoothing in supply chain and logistics forecasting. *International Journal of Production Economics*, 208, 122–139. https://doi.org/10.1016/j.ijpe.2018.11.015.

Connor, O. (1992). Exploring judgemental evidence. *International Journal of Forecasting*, 8, 15–26. https://doi.org/10.1016/0169-2070(92)90004-S.

Dalrymple, D. J. (1987). Sales forecasting practices: Results from a United States survey. *International Journal of Forecasting*, Elsevier. https://www.sciencedirect.com/science/article/pii/0169207087900318.

De Baets, S. & Harvey, N. (2018). Forecasting from time series subject to sporadic perturbations: Effectiveness of different types of forecasting support. *International Journal of Forecasting*, 34(2), 163–180. https://doi.org/10.1016/j.ijforecast.2017.09.007.

Eroglu, C. & Croxton, K. L. (2010). Biases in judgmental adjustments of statistical forecasts: The role of individual differences. *International Journal of Forecasting*, 26(1), 116–133. https://doi.org/10.1016/j.ijforecast.2009.02.005.

Fildes, R. & Goodwin, P. (2007). Against your better judgment ? How organizations can improve their use of management judgment in forecasting. *Interfaces*, 37(6) (April 2014), 570–576. https://doi.org/10.1287/inte.1070.0309.

Fildes, R. & Goodwin, P. (2021). Stability in the inefficient use of forecasting systems: A case study in a supply chain company. *International Journal of Forecasting*, 37(2), 1031–1046. https://doi.org/10.1016/j.ijforecast.2020.11.004.

Fildes, R., Goodwin, P., & Lawrence, M. (2006). The design features of forecasting support systems and their effectiveness. *Decision Support Systems*, 42(1), 351–361. https://doi.org/10.1016/j.dss.2005.01.003.

Fildes, R., Goodwin, P., & Önkal, D. (2018). Use and misuse of information in supply chain forecasting of promotion effects. *International Journal of Forecasting*, 35(1), 144–156. https://doi.org/10.1016/j.ijforecast.2017.12.006. (https://www.sciencedirect.com/science/article/pii/S0169207018300062).

Ghosh, A. & Fedorowicz, J. (2008). The role of trust in supply chain governance. *Business Process Management Journal*, 14(4), 453–470. https://doi.org/10.1108/14637150810888019.

Goodwin, P. (1996). Statistical correction of judgmental point forecasts and decisions. *Omega*, 24(5), 551–559. https://doi.org/10.1016/0305-0483(96)00028-X.

Goodwin, P., Moritz, B., & Siemsen, E. (2018). Forecast decisions. In Donohue, K., Katok, E. & Leider, S. (eds.), *The Handbook of Behavioral Operations*. https://doi.org/10.1002/9781119138341.ch12.

Klassen, R. D. & Flores, B. E. (2001). Forecasting practices of Canadian firms: Survey results and comparisons. *International Journal of Production Economics*, 70(2), 163–174. https://doi.org/10.1016/S0925-5273(00)00063-3.

Lawrence, M., Goodwin, P., O'Connor, M., & Önkal, D. (2006). Judgmental forecasting: A review of progress over the last 25 years. *International Journal of Forecasting*, 22(3), 493–518. https://doi.org/10.1016/j.ijforecast.2006.03.007.

Lee, T. (2007). Loss functions in time series forecasting. *University of California*, 1(1999), 1–14. http://www.faculty.ucr.edu/~taelee/paper/lossfunctions.pdf.

Matharage, S. T., Hewage, U., & Perera, H. N. (2020). Impact of sharing point of sales data and inventory information on bullwhip effect. *IEEE International Conference on Industrial Engineering and Engineering Management, 2020-Dec* (pp. 857–861). https://doi.org/10.1109/IEEM45057.2020.9309733.

Mehrjerdi, Y. Z. A. H. (2016). The Bullwhip effect on the VMI-supply chain management viasystem dynamics approach: The supply chain with two suppliers and one retail channel. *International Journal of Supply and Operations Management*, 3(2), 1301–1317. https://doi.org/10.22034/2016.2.05.

Önkal, D., Gönül, M. S., & Lawrence, M. (2008). Judgmental adjustments of previously adjusted forecasts. *Decision Sciences*, 39(2), 213–238. https://doi.org/10.1111/j.1540-5915.2008.00190.x.

Perera, H. N., Fahimnia, B., & Tokar, T. (2020). Inventory and ordering decisions: A systematic review on research driven through behavioral experiments. *International Journal of Operations & Production Management*, 40(7/8), 997–1039. https://doi.org/10.1108/IJOPM-05-2019-0339.

Perera, H. N., Hurley, J., Fahimnia, B., & Reisi, M. (2019). The human factor in supply chain forecasting: A systematic review. *European Journal of Operational Research*, 274(2), 574–600. https://doi.org/10.1016/j.ejor.2018.10.028.

Petropoulos, F., Fildes, R., & Goodwin, P. (2016). Do "big losses" in judgmental adjustments to statistical forecasts affect experts' behaviour? *European Journal of Operational Research*, 249(3), 842–852. https://doi.org/10.1016/j.ejor.2015.06.002.

Phillips, C. J. & Nikolopoulos, K. (2019). Forecast quality improvement with Action Research: A success story at PharmaCo. *International Journal of Forecasting*, 35(1), 129–143. https://doi.org/10.1016/j.ijforecast.2018.02.005.

Sanders, N. R. (1992). Accuracy of judgmental forecasts: A comparison. *Omega*, 20(3), 353–364. https://doi.org/10.1016/0305-0483(92)90040-E.

Sanders, N. R. & Manrodt, K. B. (1994). Forecasting practices in US corporations: Survey results. *Interfaces*, 24(2)(August 2015), 92–100. https://doi.org/https://www.jstor.org/stable/25061863.

Sanders, N. R., Manrodt, K. B., Sanders, N. R., & Manrodt, K. B. (2003). Forecasting software in practice: Use, satisfaction, and performance. *Interf,* 33(5)(June 2015), 90–93.

Tangpong, C., Hung, K. T., & Li, J. (2019). Toward an agent-system contingency theory for behavioral supply chain and industrial marketing research. *Industrial Marketing Management,* 83(January), 134–147. https://doi.org/10.1016/j.indmarman.2018.10.003.

Van den Broeke, M., De Baets, S., Vereecke, A., Baecke, P., & Vanderheyden, K. (2019). Judgmental forecast adjustments over different time horizons. *Omega,* 87, 34–45. https://doi.org/10.1016/j.omega.2018.09.008. https://www.sciencedirect.com/science/article/pii/S0305048317309246.

Webby, R., O'Connor, M., & Edmundson, B. (2005). Forecasting support systems for the incorporation of event information: An empirical investigation. *International Journal of Forecasting,* 21(3), 411–423. https://doi.org/10.1016/j.ijforecast.2004.10.005.

Zellner, M., Abbas, A. E., Budescu, D. V., & Galstyan, A. (2021). A survey of human judgement and quantitative forecasting methods. *The Royal Society Open Science,* 8(2): 201187. https://doi.org/10.1098/rsos.201187.

Chapter 11

Fuzzy Logic-Based Multi-Objective Decision-Making Model for Design Evaluation in an Open Innovation Environment

**S. Denis Ashok[*], S. Krishna[†] and
S. G. Ponnambalam[‡]**

*School of Mechanical Engineering, Department of Design and
Automation, Cyber Physical Systems Laboratory,
Vellore Institute of Technology, Vellore, Tamil Nadu, India*
[*]denisashok@vit.ac.in
[†]krishna.s@vit.ac.in
[‡]ponnambalam.g@vit.ac.in

In an open innovation environment of intersecting multidisciplinary engineering domains, it is tough to decide on suitable alternate design models which meet the conflicting, multiple objectives with constraints and uncertainty. This chapter presents the application of a fuzzy logic-based multi-objective decision-making model in order to acquire meaningful information linguistically and rank alternatives for design evaluation. The proposed method uses the linguistic ratings of domain experts to formulate fuzzy sets of multiple objectives. In the present work, a weighted logical decision function is formulated to identify the best alternative design model which meets the user preferences. A case study on mobile robot chassis design evaluation is considered to high-light the application of fuzzy logic in the multi-criteria decision-making process, and the results are presented. The proposed method can handle the linguistic information under uncertainty and it will be helpful for the decision-making process in a product development environment.

1. Introduction

Multi-criteria decision-making (MCDM) is an important field of research with potential practical applications such as weapons evaluation, aggrega-tion of market research data, technology transfer selection, and project maturity evaluation (Carlsson & Fullér, 1996). In order to help the decision-makers solve the complex decision-making problems in a systematic, con-sistent, and more productive way, there exist four distinct families of methods for MCDM, such as outranking, value and utility theory-based methods, multiple objective programming, group decision, and negotiation theory-based methods (Ali *et al.*, 2014). In spite of the significant develop-ments in computer-aided decision-making processes for product develop-ment, multi-criteria decision-making is still not explicit in handling the complexity and fuzziness of multiple criteria objectives related to design evaluation. Also, there is a tendency toward information loss due to subjec-tivity and expectation of experts during the rating of alternative design solutions. A typical MCDM method needs to have the capability of combin-ing quantitative and qualitative criteria to select the best alternatives in the presence of uncertainty, risk, and incomplete information (Blockley, 1982). Hence, there is a need to develop a suitable computer-assisted decision-making model which handles uncertainty in evaluating the alternate solu-tions during the product development and design evaluation process.

Due to applications in various domains, many approaches have been developed for Multiple Criteria Decision-Making such as SAW, AHP, TOPSIS, PROMETHEE, ELECTRE, TODIM, LINMAP, COPRAS, VIKOR, ARAS, MOORA, MULTIMOORA, WASPAS, SWARA, KEMIRA, EDAS, and FARE. Several researchers advocate the use of fuzzy logic to deal with the problems related to prioritization and decision-making, and there are several interesting results on group deci-sion-making with the help of fuzzy set theory. Fuzzy set presents flexibil-ity to deal with the vagueness and no specificity inherent in human formulation of preferences, constraints, and goals (Zadeh, 1965). Aggregating methods based on group and individual preferences are presented in multi-criteria decision-making applications (Nakamura, 1986). Fuzzy ranking methods are presented and their application for dealing with multiple criteria is presented (Yuan, 1991). However, these methods are difficult to implement due to the computational complexity. Li (1999) proposed a method using an a-level weighted F-preference relation and an efficient fuzzy model for a multiple criteria decision problem in a fuzzy environment (Li, 1999). In this approach, the performance ratings and the criteria are defined imprecisely and represented by fuzzy numbers. As the criteria have varying importance and preferences, there is a need for weighting the criteria (Dulmin & Mininno, 2003). The AHP approach was developed in the early 1970s in response to military contingency planning, scarce resource allocation, and the need for political participation in disarmament agreements (Saaty, 1977). An empirical study was carried out to evaluate the importance of the supplier selection and assessment criteria of American manufacturing companies (Kannan & Tan, 2002). Lee *et al.* proposed a methodology for identifying managerial criteria using information derived from the supplier selection processes and supplier management process (Lee *et al.*, 2001). Differences in decision criteria are analyzed in a multi-criteria decision-making problem (Pearson & Ellram, 1995). Different weights are assigned to decision criteria and its perceived importance on the best selection is analyzed (Verma & Pullman, 1998). Fuzzy sets and their applications for dealing with uncertainty and vagueness for various industrial applications are highlighted (Klir & Folger, 1988; Klir & Yuan, 1995). A decision table with analytic network process and fuzzy inputs is developed (Ramik & Vlach, 2002).

An open innovation approach for based product development brings a collaborative model involving companies, research centers, universities and experts. Design decision-making and product evaluation are challenging tasks in an open innovation environment due to the associated risk, limited time, and resources. Recent applications of different multi-criteria decision-making problems are listed in Table 1.

It is noted that multi-criteria decision-making for design evaluation in the specific context of mobile robotics has gained little attention. Also, the weighted objectives and user preferences in decision-making require complex calculations. In order to develop a robust design evaluation approach with multiple criteria in the presence of uncertainty, this chapter presents a fuzzy logic-based evaluation approach to support a computer-assisted decision-making model in product design evaluation. First, alternate design models are organized and necessary objectives for the evaluation are gathered by domain experts.

Table 1.   Recent application of fuzzy sets for MCDM problems.

Application domain	Authors
Pattern recognition problems	Meng & Chen (2014)
Bridge risk assessment	Shen *et al.* (2016)
Ranking investment alternatives	Zavadskas *et al.* (2015) and Hashemi *et al.* (2016) Zavadskas *et al.* (2014) and Hajiagha *et al.* (2013b)
Air-condition system selection	Hashemkhani Zolfani *et al.* (2015) and Liu & Wang (2007) Lin *et al.* (2007)
Tourism management	Chu & Guo (2015)
Revitalization of buildings	Zavadskas *et al.* (2014)
Customer satisfaction determination	Hajiagha *et al.* (2013a)
Personnel selection	Wan *et al.* (2013)
Medical diagnosis	Chen (2015) and De *et al.* (2001)
Shareholdings and stocks	Wang (2021)

Source: Stanujkic et al. (2017).

Linguistic ratings for the design alternatives are gathered for each criterion from the decision-makers. Ordered fuzzy sets are formulated to extract a reasonable understanding of linguistic ratings from the decision-makers. Weighted evaluation of the performance ratings for each criterion is calculated using the logical operation of fuzzy sets.

This chapter is organized as follows. In Section 2, terminologies related to multi-criteria decision-making model are introduced, and the basics of fuzzy sets are discussed in Section 3. Section 4 presents the steps involved in the proposed method for the multi-criteria decision-making model. A case study on design evaluation of mobile robot configuration using the proposed fuzzy logic-based multi-criteria decision-making model is explained with the results in Section 5.

2. Terminologies in Multi-Criteria Decision-Making

In this work, application of MCDM is used to determine the best design which fits a set of criteria based on the ratings of the decision-makers. Here, there will be multiple criteria which will be usually conflicting. This section presents the terminology related to MCDM.

2.1. *Alternatives*

Alternatives refer to the available design models which will be applied for the ranked selection.

2.2. *Criteria*

Criteria refer to the quality of the alternatives which needs to be satisfied to meet the specific requirements. The MCDM is associated with a set of criteria or attributes that will impact the selection of the alternatives. The typical criteria for the design alternatives can be based on the perspectives of manufacturing cost, innovation, simplicity, assembly time, and feasibility.

2.3. *Weights*

Weights indicate relative importance of the criteria which define the preference and requirement. The weights are standardized, and the sum of them is one.

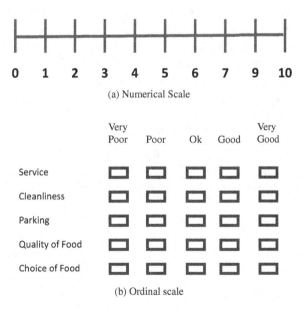

Figure 1.   Illustration of typical numerical and ordinal scales for rating. (a) Numerical Scale; (b) Ordinal scale.

2.4. Decision-Makers (DMs)

Domain experts will rank and provide the weight for each criterion of the alternative based on their knowledge and expertise. Ordinal and numerical scales will be used for the decision-making process as shown in Figure 1.

2.5. Decision matrix

It is matrix which will be used for making the objective decisions from the given alternate options. A typical MCDM can be described by the decision matrix D which includes the "*m*" alternatives and "*n*" criteria of the design as shown in the following:

$$D = r_{ij} = \begin{bmatrix} r_{11} & r_{12} & \cdots\cdots & r_{1n} \\ r_{21} & r_{12} & \cdots\cdots & r_{2n} \\ \cdots & \cdots\cdots\cdots & \\ r_{m1} & r_{12} & \cdots\cdots & r_{mn} \end{bmatrix} \tag{1}$$

A set of "*m*" design alternatives is obtained as computer-aided design models $\{A_1, A_2, \ldots, A_m\}$ and "*n*" criteria $\{C_1, C_2, \ldots, C_n\}$ are formulated for rating by a group of k decision-makers (DMs) $\{DM_1, DM_2, \ldots, DM_k\}$. The weights for the attributes are considered as $\{w_1, w_2, \ldots, w_n\}$. Table 2 indicates the decision matrix with ratings "r_{mn}" by the decision-maker involving "*m*" alternatives and "*n*" criteria and weights.

Table 2.   Decision matrix as a table.

Alternatives	C_1 W_1	C_2 W_2	C_3 W_3	...	C_n W_n
A_1	r_{11}	r_{12}	r_{13}	...	r_{1n}
A_2	r_{21}	r_{22}	r_{23}	...	r_{2n}
A_3	r_{31}	r_{32}	r_{33}	...	r_{3n}
A_4	r_{41}	r_{42}	r_{43}	...	r_{4n}
...	...	...	...	...	...
...	...	...	...	...	...
A_m	r_{m1}	r_{m2}	r_{m3}	...	r_{mn}

To rate the design alternatives, linguistic terms are used by the decision-makers as they provide the natural mode of representing human knowledge and also a way of dealing with uncertainty or vague information.

3. Basics of Fuzzy Sets and Fuzzy Numbers

In this section, a fuzzy set framework is presented to deal with the uncertainty in the linguistic ratings of the criteria by the decision-makers. In this work, a discrete set of linear ordered linguistic variables from the worst one to the best one is converted into a numerical order scale using fuzzy sets and fuzzy numbers.

As the vague linguistic variable has no sharp boundary, a fuzzy set is used to handle the imprecision that characterizes much of human reasoning in the predefined space using membership functions. A fuzzy set assigns levels of membership μ in a range [0, 1] for each element of x in a set A in a universe U. Fuzzy set A may be written by the set of pairs as follows:

$$A = \{(x, \mu_A(x)), x \in X\} \tag{2}$$

The value $\mu_A(x)$ characterizes the grade of membership of x in A.

The mathematical expressions for building the triangular, trapezoidal, and Gaussian membership functions are given below. The triangular function is defined by the parameters, a, b, m where $a < m < b$ is given by the following Equation (3):

$$\mu_A(x) = \begin{cases} 0, & (x < a) \text{ or } (x < d) \\ \frac{x-a}{b-a} & a \le x \le b \\ 1, & b \le x \le c \\ \frac{d-x}{d-c} & c \le x \le d \end{cases} \tag{3}$$

The triangular, trapezoidal, and Gaussian functions are commonly used to build membership functions for the fuzzy sets as illustrated in Figure 2. It graphically represents the fuzzy sets.

In Equation (3), a refers to the lower bound, b is the peak point, and c the upper bound of the triangular function.

As shown in Figure 2(b), a typical trapezoidal function is defined by a lower bound a, an upper bound d, a lower support bound b, and

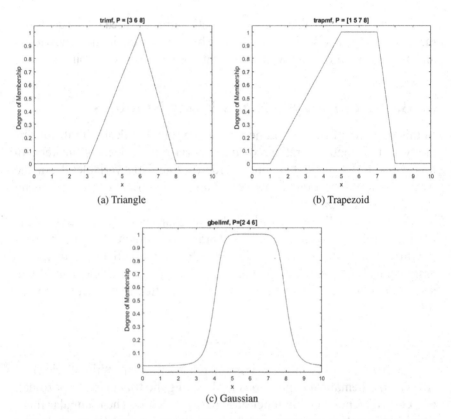

Figure 2.   Typical membership functions. (a) Triangle; (b) Trapezoid; (c) Gaussian.

an upper support bound c, where $a < b < c < d$ using the following Equation (4):

$$\mu_A(x) = \begin{cases} 0, & x \le a \\ \frac{x-a}{m-a} & a < x \le m \\ \frac{b-x}{b-m} & m < x < b \\ 0, & x \ge b \end{cases} \tag{4}$$

A Gaussian function **is** defined by a central value m and a standard deviation $k > 0$ as given by the following Equation (5):

$$\mu_A(\mathrm{x}) = e^{-\frac{(x-m)^2}{2k^2}} \tag{5}$$

The higher value of k increases the spread of the Gaussian function.

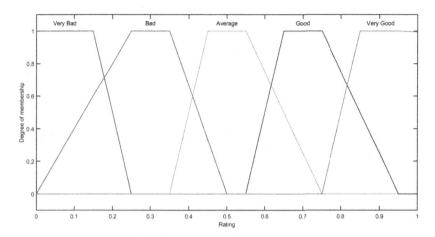

Figure 3. Trapezoid fuzzy sets.

Using the fuzzy sets and membership functions, fuzzy numbers are formulated to transform the qualitative linguistic variables into a numerical scale.

The uncertainty associated with the ratings of the decision-maker can be expressed with more spread of the fuzzy number. Here, the scale of fuzzy numbers is constructed on the basis of a particular situation. The fuzzy number also depicts the categorization of linguistic variables in a suitable numerical scale.

Typically, the linguistic variables are labeled as very small, small, medium, large, and very large, and can be represented by trapezoid fuzzy numbers as shown in Figure 3.

4. Fuzzy Number-Based Multi-Criteria Decision-Making Model

In the present work, a fuzzy number-based multi-objective decision-making model is proposed for the selection of design alternatives for product development applications and it has following steps.

4.1. *Formulate the number of evaluation criteria and the design alternatives*

A universe of "n" alternative design models of the product, $a = \{a_1, a_2,..., a_n\}$, and a set of "$r$" multiple objectives, $O = \{O_1, O_2,..., O_r\}$, is formulated for the evaluation of design models.

In this work, domain experts assign linguistic ratings for the multiple objectives of the alternate design models, and fuzzy sets are formulated using suitable membership values as denoted by $\mu_{Oi}(a)$, for which the alternative design model "*a*" needs to satisfy criteria specified for this objective.

4.2. Perform linguistic rating of the design alternative using domain experts and decision-makers

Linguistic ratings of the design alternatives are by the decision-makers in a order scale and fuzzy sets are formulated for encoding the ordinal features of the linguistic ratings. The order scale of the linguistic terms used for rating the alternate design is given by the following set:

$$OS = \{\text{Very Bad, Bad, Average, Good, Very good}\} \qquad (6)$$

4.3. Develop fuzzy sets and fuzzy numbers for transforming the linguistic variable in a numerical order scale

Fuzzy sets and Fuzzy numbers are formulated to transform the given linguistic variable to handle the uncertainty of information on ratings given by the decision-maker. Table 3 shows the order scale and the corresponding fuzzy number used in the present work.

It involves fuzzification using shapes and the number of the membership functions (mf) for the input variables.

Table 3.   Linguistic variables and fuzzy number scale.

Linguistic variable	Order scale	Fuzzy number
Very bad	VB	0.2
Bad	B	0.4
Average	A	0.6
Good	G	0.8
Very good	VG	1.0

A membership function defines the extent to which each value of a numerical variable belongs to specified linguistic variables as given by the following equation:

$$\hat{O}_r = \frac{\mu_1}{a_1} + \frac{\mu_2}{a_2} + \cdots + \frac{\mu_n}{a_n} \tag{7}$$

Here, $\mu_1, \ldots, \mu_n$ represents the membership value for the given linguistic rating of each alternate design model as given by Equation (8):

$$a = \{a_1, a_2, \ldots, a_n\} \tag{8}$$

Triangular, Trapezoid, and Gaussian membership functions can be used for representing the fuzzy numbers in a suitable numerical scale.

4.4. *Select suitable weights for the criterion to describe the relative importance*

Suitable weights are assigned to each objectives for expressing the relative importance and preference of the decision-maker. A weighted decision-making function is formulated using set of weights $\{W_r\}$ where $i = 1, 2, \ldots, r$. for the multiple objectives and it is represented as the intersection of r-tuples, denoted as $M(O_i, W_i)$ as given in the following:

$$D_w = M(O_1, W_1) \cap M(O_2, W_2) \cap \cdots \cap M(O_r, W_r) \tag{9}$$

4.5. *Apply intersection on the fuzzy numbers of the performance ratings with respect to all the criteria for each alternative as aggregated evaluation*

A logical decision function is formulated that simultaneously satisfies multiple objectives $O = \{O_1, O_2, \ldots, O_r\}$ by using the logical intersection of all the objective fuzzy sets as given in the following:

$$D = O_1 \cap O_2 \cap \cdots \cap O_r \tag{10}$$

The grade of membership for the decision function, D, of each alternative design models "a" is given as follows:

$$\mu_D(a) = min[\mu_{O1}(a), \mu_{O2}(a), \ldots, \mu_{Or}(a)] \tag{11}$$

The optimum decision for the alternative $a*$, which meets the multiples objectives, is obtained by the following:

$$\mu_D(a*) = max\,(\mu_D(a)),\ a \in A \qquad (12)$$

As there is a unique relationship between a weight and its associated objective, the weighted decision-making model is obtained using the classical implication operation of weights and the corresponding objectives as given by the following:

$$D_w = \bigcap_{i=1}^{r}(\bar{W}_i \cup O_i) \qquad (13)$$

and the optimum solution, $a*$, is the alternative that maximizes the decision function D.

5. Application of Proposed Method for Design Evaluation of Mobile Robot Chassis: A Case Study

This work presents a case study on evaluation of a design model of a mobile robot configuration in an open innovation environment. The chassis of the mobile robot is an important component which holds the major elements of the mobile robot and provides effective weight distribution. Three design models $a = \{a_1, a_2,..., a_3\}$ are constructed in the computer-aided design environment and the best design alternative is evaluated using the proposed weighted fuzzy logic-based multi-criteria decision-making model. Figure 4 shows the computer-aided design models of the alternate mobile robot.

In order to promote innovation in the chassis design of the mobile robot, five objectives $O = \{O_1, O_2,..., O_5\}$ are formulated such as Increased Novelty, Reduced Manufacturing Cost, Minimum Assembly time, Reduced design Complexity, and Increased Feasibility for house manufacturing.

5.1. *Linguistic ratings of mobile robot chassis*

Domain experts with interdisciplinary engineering knowledge in design, manufacturing, mechatronics are involved for providing the ratings to the design models of mobile robot chassis.

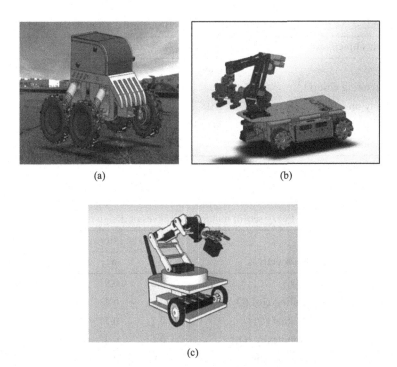

Figure 4.   Design models of mobile robot chassis. (a) All terrain design; (b) Rectangular chassis design; (c) Square chassis design.

Predefined linguistic terms are used for rating the design models based on multiple objectives such as increased novelty, reduced manufacturing cost, minimum assembly time, reduced design complexity, and increased feasibility for house manufacturing as given in Table 4. It is noted that the second design model a_2 is found to be better in novelty with reduced manufacturing cost and design complexity due its simplified construction as compared to the other design alternatives.

These linguistic ratings are in the natural language and they are often considered as an important form of information, but it is tough to interpret in the computer-assisted decision-making process. In order to extract meaningful information from the linguistic ratings, fuzzy sets for each chassis design of mobile robot are formulated using the membership functions.

Table 4.   Linguistic ratings of multiple objectives for the given design models.

Multiple objectives	a_1	a_2	a_3
Novelty (O_1)	Good	Very good	Good
Manufacturing cost (O_2)	Bad	Good	Average
Assembly time (O_3)	Bad	Average	Very bad
Design complexity (O_4)	Very bad	Good	Bad
Manufacturing feasibility (O_5)	Good	Average	Very good

Table 5.   Fuzzy sets of multiple objectives of design models.

Multiple objectives	Design alternatives		
	a_1	a_2	a_3
Novelty (O_1)	0.6	0.9	0.7
Manufacturing cost (O_2)	0.9	0.4	0.6
Assembly time (O_3)	0.5	0.7	0.3
Design complexity (O_4)	0.1	0.9	0.5
Manufacturing feasibility (O_5)	0.9	0.5	0.1

5.2. *Formulation of fuzzy sets*

Fuzzy sets are used for handling uncertainty in the linguistic ratings provided by the domain experts, and it is mentioned in Table 5.

From Table 5, it can be seen that the values range from 0 to 1, which quantifies the characteristics of multiple objectives numerically. Using the predefined geometric membership functions such as triangle and Gaussian, membership grades are extracted and assigned for each design alternative. Here, the domain expertise and experience are used to construct membership functions which transform the linguistic ratings into a fuzzy single ton.

Figure 5 shows the graphical illustration of fuzzy numbers for different objectives of the design models of the robot chassis.

It is tough to select the best design model which meets the given objectives. It requires logical decision-making using the developed fuzzy sets.

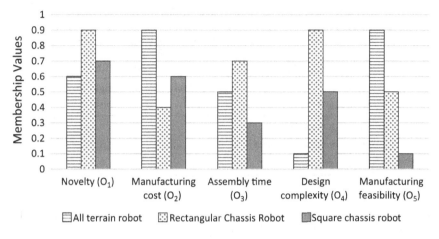

Figure 5. Fuzzy numbers for the multiple objectives.

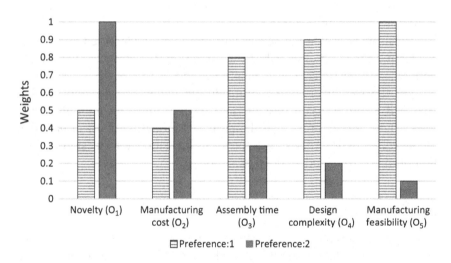

Figure 6. Different weights for the multiple objectives of the decision-making model.

5.3. *Weighted logical decision-making function*

In order to improve decision-making, user-defined preferences are considered and weights are assigned accordingly to the multiple objectives. Figure 6 shows the different weights considered in the present work to highlight the preference of the decision-maker in selecting the mobile robot chassis. From Figure 6, the maximum value of 0.9 indicates the

Table 6.   Evaluation of design models using weighted decision-making function.

Design alternatives	Weighted decision-making function	
	Weight for case: 1	Weight for case: 2
Design model: 1	0.3	0.4
Design model: 2	0.6	0.5
Design model: 3	0.4	0.7

preference for the manufacturing cost and it shows the preference for the novelty of the mobile robot chassis design.

Based on the given weights for the multiple objectives, the design models are evaluated using the proposed decision-making model, and the results are presented in Table 6.

It can be seen that the second design model is identified as the best alternative for the mobile robot chassis design which meets the weight preference of the objectives as mentioned in Case 1.

For the weight in Case 2, third design model for the mobile robot chassis is identified as the best alternative which meets the multiple objectives. These results prove that the proposed model is robust in handling the weighted multi-criteria objectives and it is suitable for design evaluation of innovative product developments.

6. Conclusions

This chapter presented an application of a fuzzy set and fuzzy number for a multi-criteria decision-making model for the design evaluation of a mobile robot. The proposed method is suitable in handling the conflicting, multiple objectives with constraints and uncertainty in an open innovation environment of intersecting multi-disciplinary engineering domains. Mobile robot chassis design models are evaluated using the proposed method for multiple objectives such as Increased Novelty, Reduced Manufacturing Cost, Minimum Assembly time, reduced design Complexity, and Increased Feasibility for house manufacturing.

Fuzzy numbers are used to extract the linguistic ratings of the design models for the given objectives, and a logical decision-making model is used to arrive at the best alternative which meets the specified objectives.

In order to provide the reasonable decision-making model with user preferences on the multiple's objectives, weights are assigned to the objectives for selecting the best alternatives of the design model. Proposed model will be useful in developing a computer assisted decision-making process for product selection and evaluation of design models. The proposed method can be extended by involving institutional fuzzy sets to improve the design evaluation in a product development environment.

Acknowledgments

This research work was funded by the Science Engineering Research Board, New Delhi, under the Teaching Associateship for the Research Excellence Scheme (TAR/2018/001123) to carry out a collaborative project by the Teaching Learning Centre, Department of Mechanical Engineering, Indian Institute of Information Technology, Design, Manufacturing, Kanchipuram, Tamil Nadu, India, and Cyber Physical Systems Lab, Vellore Institute of Technology, Vellore, Tamil Nadu, India.

References

Ali, R., Moncef, T., & Khaled, G. (2014). Fuzzy model for multicriteria decision making. In *2014 Information and Communication Technologies Innovation and Application (ICTIA)*.

Blockley, D. (1982). Comments on "model uncertainty in structural reliability" by Ove Ditlevsen. *Structural Safety*, 1(3), 233–235.

Carlsson, C. & Fullér, R. (1996). Fuzzy multiple criteria decision making: Recent developments. *Fuzzy Sets and Systems*, 78(2), 139–153.

Chen, T.-Y. (2015). The inclusion-based TOPSIS method with interval-valued intuitionistic fuzzy sets for multiple criteria group decision making. *Applied Soft Computing*, 26, 57–73.

Chu, C.-H. & Guo, Y.-J. (2015). Developing similarity-based IPA under intuitionistic fuzzy sets to assess leisure bikeways. *Tourism Management*, 47, 47–57.

De, S. K., Biswas, R., & Roy, A. R. (2001). An application of intuitionistic fuzzy sets in medical diagnosis. *Fuzzy Sets and Systems*, 117(2), 209–213.

Dulmin, R. & Mininno, V. (2003). Supplier selection using a multi-criteria decision aid method. *Journal of Purchasing and Supply Management*, 9(4), 177–187.

Hajiagha, S. H. R., Akrami, H., Zavadskas, E. K., & Hashemi, S. S. (2013a). An intuitionistic fuzzy data envelopment analysis for efficiency evaluation under uncertainty: Case of a finance and credit institution. *E & M Ekonomie A Management*, 16(1), 128–137.

Hajiagha, S. H. R., Hashemi, S. S., & Zavadskas, E. K. (2013b). A complex proportional assessment method for group decision making in an interval-valued intuitionistic fuzzy environment. *Technological and Economic Development of Economy*, 19(1), 22–37.

Hashemi, S. S., Hajiagha, S. H. R., Zavadskas, E. K., & Mahdiraji, H. A. (2016). Multicriteria group decision making with ELECTRE III method based on interval-valued intuitionistic fuzzy information. *Applied Mathematical Modelling*, 40(2), 1554–1564.

Hashemkhani Zolfani, S., Sedaghat, M., Maknoon, R., & Zavadskas, E. K. (2015). Sustainable tourism: A comprehensive literature review on frameworks and applications. *Economic Research-Ekonomska Istraživanja*, 28(1), 1–30.

Kannan, V. R. & Tan, K. C. (2002). Supplier selection and assessment: Their impact on business performance. *The Journal of Supply Chain Management*, 38(4), 11–21.

Klir, G. J. & Folger, T. A. (1988). *Fuzzy Sets, Uncertainty, and Information*. Englewood Cliffs, N.J.: Prentice Hall.

Klir, G. J. & Yuan, B. (1995). *Fuzzy Sets and Fuzzy Logic: Theory and Applications*. Upper Saddle River, N.J.: Prentice Hall.

Lee, E.-K., Ha, S., & Kim, S.-K. (2001). Supplier selection and management system considering relationships in supply chain management. *IEEE Transactions on Engineering Management*, 48(3), 307–318.

Li, R.-J. (1999). Fuzzy method in group decision making. *Computers & Mathematics with Applications*, 38(1), 91–101.

Lin, L., Yuan, X.-H., & Xia, Z.-Q. (2007). Multicriteria fuzzy decision-making methods based on intuitionistic fuzzy sets. *Journal of Computer and System Sciences*, 73(1), 84–88.

Liu, H.-W. & Wang, G.-J. (2007). Multi-criteria decision-making methods based on intuitionistic fuzzy sets. *European Journal of Operational Research*, 179(1), 220–233.

Meng, F. & Chen, X. (2014). Entropy and similarity measure of Atanassov's intuitionistic fuzzy sets and their application to pattern recognition based on fuzzy measures. *Pattern Analysis and Applications*, 19(1), 11–20.

Nakamura, K. (1986). Preference relations on a set of fuzzy utilities as a basis for decision making. *Fuzzy Sets and Systems*, 20(2), 147–162.

Pearson, J. M. & Ellram, L. M. (1995). Supplier selection and evaluation in small versus large electronics firms. *Journal of Small Business Management*, 33(4), 53–65.

Ramik, J. & Vlach, M. (2002). Fuzzy mathematical programming: A unified approach based on fuzzy relations. *Fuzzy Optimization and Decision Making*, 1(4), 335–346.

Saaty, T. L. (1977). A scaling method for priorities in hierarchical structures. *Journal of Mathematical Psychology*, 15(3), 234–281.

Shen, F., Xu, J., & Xu, Z. (2016). An outranking sorting method for multi-criteria group decision making using intuitionistic fuzzy sets. *Information Sciences*, 334–335, 338–353.

Stanujkic, D., Zavadskas, E. K., Karabasevic, D., Urosevic, S., & Maksimovic, M. (2017). An approach for evaluating website quality in hotel industry based on triangular intuitionistic fuzzy numbers. *Informatica*, 28(4), 725–748.

Verma, R. & Pullman, M. E. (1998). An analysis of the supplier selection process. *Omega*, 26(6), 739–750.

Wan, S.-P., Wang, Q.-Y., & Dong, J.-Y. (2013). The extended VIKOR method for multi-attribute group decision making with triangular intuitionistic fuzzy numbers. *Knowledge-Based Systems*, 52, 65–77.

Wang, J.-W. (2021). Multi-criteria decision-making method based on a weighted 2-tuple fuzzy linguistic representation model. *International Journal of Information Technology & Decision Making*, 20(2), 619–634.

Yuan, Y. (1991). Criteria for evaluating fuzzy ranking methods. *Fuzzy Sets and Systems*, 43(2), 139–157.

Zadeh, L. A. (1965). Fuzzy sets. *Information and Control*, 8(3), 338–353.

Zavadskas, E. K., Antucheviciene, J., Razavi Hajiagha, S. H., & Hashemi, S. S. (2014). Extension of weighted aggregated sum product assessment with interval-valued intuitionistic fuzzy numbers (WASPAS-IVIF). *Applied Soft Computing*, 24, 1013–1021.

Zavadskas, E. K., Antucheviciene, J., Razavi Hajiagha, S. H., & Hashemi, S. S. (2015). The interval-valued intuitionistic Fuzzy MULTIMOORA method for group decision making in engineering. *Mathematical Problems in Engineering*, 2015, 1–13.

https://doi.org/10.1142/9781786349989_0012

Chapter 12

A Sentiment-Based Approach for Innovative Product Sales Forecasting

M. A. Premachandra[*,†,‡,§], **P. T. Ranil S. Sugathadasa**[*,†,‡,¶],
Amila Thibbotuwa[*,†,‡,∥] **and H. Niles Perera**[*,†,‡,**]

[*]*Faculty of Engineering, Department of Transport and Logistics
Management, Moratuwa, Sri Lanka*
[†]*Center for Supply Chain, Operations and Logistics Optimization,
University of Moratuwa, Katubedda, Sri Lanka*
[‡]*Professor H.Y. Ranjit Perera Institute for Applied Research,
Nugegoda, Sri Lanka*
[§]*asankapremachandra95@gmail.com*
[¶]*ranil@uoml.lk*
[∥]*amilat@uom.lk*
[**]*hniles@uom.lk*

Purpose: Established information can be detected, obtained, combined, and deployed to solve new problem domains using data analytics technology, which can speed up the innovation process. External information usage makes rigorous contribution to innovation. Therefore, consumer behavior analysis has become an important instrument for an innovation-driven market, and there are various innovative analytical approaches to predict products' sale performances. Therefore, this chapter identifies the impact of electronic word of mouth (eWOM) attributes to predict user-sensitive product sales performance.

Design/methodology/approach: The Valence Aware Dictionary for sentiment Reasoning (VADER) text analyzing method was deployed to properly identify consumer opinions. The prophet forecasting model was developed as a basic forecasting model and improved for accuracy using different classified variables of eWOM.

Findings: The relative improvement of the MAPE value was 10%, 13%, and 42% with sentiment, sentiment with volume percentage, and sentiment with volume use as an extra regressor to base model, respectively. The predictive capability of interplay variables performs a more integral role than those regressors used as an individual regressor to the model. A new eWOM combination was identified which is a novel and innovative contribution to literature and practice. This method reveals data that can assist leaders in discovering new ways to be more innovative.

1. Introduction

The development of new solutions that need to fulfill consumer requirements or generate value for society is considered innovation (Andreassen & Streukens, 2009). Innovation can either be closed innovation (within a company) or opened innovation (with external parties). External party involvement such as consumer characteristics is becoming vital to product development for an e-commerce platform. Innovation can be applied to all product stages, from product identification to receiving feedback (within all products' stages) (Grützmann *et al.*, 2019). Therefore, consumer behavior identification through various innovative analytics is important in product development and it generates value to the society.

Advances in innovation and digitization enhance analytical capabilities, which is described as the ability to process, analyze, and transform data to identify trends, support decision-making, and find useful insights. Analytics can speed up the pace at which various concepts can be combined in new problem domains by gaining access to a wide range of information both inside and outside the company. Therefore, innovative analytic activities help to gain important values for an organization (Wu *et al.*, 2019).

Over the last decade, the way people purchase and sell products and services has changed due to the internet. Customer's shopping experiences are being transformed through online retail or e-commerce portals. E-tailers distinguish themselves by delivering innovative service offerings to consumers (Deloitte, 2015). Because organizations are competing for customer value creation, innovation is required to provide the proper value to customers. Product innovation entails developing new products, improving existing ones, or creates different presenting ways in order to grow and improve their utility. Therefore, products which use innovative approaches to reach consumers are considered innovative products (Yu *et al.*, 2015). Amazon.com and Alibaba.com are two successful examples. E-retailers are utilizing different features to identify consumer behavior, such as product rating and opinion sharing (word of mouth). Those attributes and reviews could be used to make sales, and they provide valuable information about the products to customers. Then, products can quickly be designed and changed due to the consumer requirements. Consumer behavior identification and managing relationships in modern e-commerce services are becoming important (Subramanian *et al.*, 2016). The quantitative and qualitative attributes of the consumer behavior are being used to predict sales through various analytical approaches. Therefore, innovative analytics is emerging in all categories as a market for digital commerce growth, and e-platforms are becoming a reliable source to make purchases.

Word of mouth (WOM) is a communication method that is used to transfer information. The electronic word of mouth (eWOM) user information sharing, which is based on reviews and comments on internet platforms, online-based promotions, e-mails, virtual conversations, websites, blogs, and social media, has added value to user identification. It has generated various innovative approaches due to characteristics of accessibility, measurability, scalability, and availability of eWOM compared to traditional methods (Cheung & Thadani, 2012). Therefore, Innovative analytics should be focused on to identify the contribution of eWOM attributes to future sales predictions.

Different approaches have been implemented which relates to e-platforms to properly identify consumer behavior. One of the most effective ways to generate internet word-of-mouth or consumer opinion is through user-generated content (UGC). It has a significant impact on the internet market since online customers frequently study customer-generated reviews while making purchasing decisions. Given the importance of online UGC, an increasing number of online communities allow users to share their thoughts and product suggestions in order to attract new customers and keep the existing ones. Consumer demand identification by using user-generated content (UGC) leads to accurate sales predictions (Duan *et al.*, 2008a).

UGC can be divided as quantitative and qualitative. Qualitative aspects of consumer opinions are an emerging development in innovative approaches to accurately forecast trends rather than analyzing quantitative aspects (Fang & Zhan, 2015). This provides significant consumer insides to understand the future trends. Therefore, innovative approaches should be generated to identify a change in consumer qualitative characteristics for a better online shopping experience.

The research has focused on using unstructured textual data to extract valuable information that cannot be explained from numeric ratings. Beyond the effect of numerical elements, the impact of review content in which online customers express their thoughts must be investigated. Therefore, quantitative characteristics were measured by using sentiment analysis. Online review sentiment can provide information that quantitative ratings are unable to fully convey. It presents the reviewer's positive, negative, or neutral opinions, or a combination of them, in order to influence customers' cognitive assessments. For both positive and negative sentiments in each review, the strength, which shows the levels of orientation, is measured and considered as sentiment value.

The qualitative and quantitative characteristics have analyzed their impact on sales performances. The VADER intensify text analyzing technique was used to measure the sentiment value, and the prophet forecasting model was investigated to identify the impact of consumer opinions on products. The prophet forecasting model was used as a basic forecasting model and model accuracy was improved by using UGC. In this chapter, besides examining the sentiment of UGC, the importance of examining sentiment values with interplay variables for sales performances was analyzed.

According to research findings, first, the analysis helps e-commerce-based decision-makers predict their innovative product sales by using real-time information. The influence of predictive variables was illustrated, and therefore, they will properly contribute to work on SCM activities and increase their supply chain efficiency with real-time information. Second, it helps to understand consumer behavior with the use of qualitative and quantitative characteristics of the UGC in the e-commerce. The importance of variables in different situations was discussed. It allows innovation decision-makers/managers to give priority to specific consumer characteristics when identifying user behavior. Providing an e-platform for consumer will facilitate more benefits for innovative product selling and marketing programs. Finally, it can be justified that the usability of sentiment analysis with the prophet forecasting model shows a better outcome that relates to innovative consumer-sensitive products.

In the following sections, this will be further explained: Section 2 provides a comprehensive background study on the impact of consumer behavior on decision-making and sales predictions. Section 3, the methodology, elaborates on the importance of selecting text analysis and variables. Section 4 presents the findings, discussion, conclusion, and future research direction.

2. Literature Review

Users tend to search for product information through digital media before making purchases and this has led to the growth of online retail shops such as Amazon or Alibaba. These retail shops are getting a stronger market position among consumers (Hou *et al.*, 2017). Due to the connectivity of consumers, their thoughts and insights can be used to improve future sales. Social media contributes massively to eWOM. It is easy to collect information from social media and understand market uncertainties with many opportunities for innovation (Schaer *et al.*, 2018). The UGC has a significant role and impact in changing others' attitudes toward innovative product performances either positively or negatively (Themistocleous *et al.*, 2004). Therefore, user opinion is critical for SCM activities.

2.1. *Importance of UGC for sentiment approach*

User opinion is a vital factor for product purchasing. WOM is deemed to be an information transaction between consumers, and it can influence consumer behavior and attitude toward the product or service. It is justified that 50% of service provider replacement is based on WOM (Keaveney, 1995) and over two-thirds of consumers consider others' opinions (Hennig-Thurau *et al.*, 2004). The WOM is categorized as negative word of mouth (NWOM) and positive word of mouth (PWOM) product opinion. The PWOM helps to encourage product or brand purchase, while NWOM discourages brand or product purchase, and the probability of a purchase differs based on the impact of WOM incidence (East *et al.*, 2008). For example, if a person receives evidence that confirms a product is unreliable, it is considered as more valuable information than the evidence that confirms the product is reliable. Because reliability is considered as the default condition of innovative products, negative information is considered as a more powerful credibility source (Rozin & Royzman, 2001), even though it might be surprising and suspicious. Therefore, negative information has a dominant impact on decision-making.

UGC has significantly more impact than other channels. The innovation analytical model is developed to argue about qualitative aspects of reviews and it provides a new communication component in marketing communication (Chen & Xie, 2008). It is discovered that consumer WOM is still believable in spite of the firms' advertising intent about innovative products (Mayzlin, 2006; Ramanathan *et al.*, 2017). The UGC is more powerful in innovation analytics than traditional marketing and marketer-generated content analysis on innovative product purchases (Campbell *et al.*, 2014). Therefore, measuring the negative and positive impacts of usergenerated content especially related to text-based opinion is important in e-commerce.

2.2. *UGC-based innovative analytics for sales prediction*

Consumer real-time behavior identification in dynamic situations leads to better results to the organization. At present, innovative businesses are maintaining or tend to use at least one social media platform/ user platform to reach and collect real-time consumer data. The number of channels for consumer service is associated with a higher reputation score and PWOM makes a massive contribution to enhance reputation (Guo *et al.*, 2020). Two different methods are developed with and without UGC and NOWM innovative analytics model has proved that statistically significant improvement in sales forecasting with a relative accuracy (Cui *et al.*, 2017). UGC is used to optimize operational decisions by using external ego. Short-period forecasting is a more significant element in the apparel industry due to the dynamic market environment. According to Iftikhar & Khan (2020), there is a correlation between consumer opinion and actual sales. They conclude that insights of rising adopters relate to positive sentiments and negative sentiment reduces the number of users in the e-commerce services. Automobile sales forecasting was conducted using a combination of sentiment and 242 macroeconomic indicators, browsing numbers, ratings, and approval numbers to calculate WOM impact. It shows significant improvement in product sales when implementing WOM for forecasting (Zhang *et al.*, 2020). The sentiment value integrates with the imitation coefficient in Norton and bass models. It improves the predicting accuracy of sales with UGC (Fan *et al.*, 2017). The WOM and UGC have a considerable impact on the aggregated level of economic outcomes. The field relates to stock market sales predictions (Antweiler & Frank, 2004; Das & Chen, 2007; Tirunillai & Tellis, 2012), box office sales forecasting (Liu *et al.*, 2007), judgment, and decision-making (Singh *et al.*, 2018). UGC has been used in several fields such as in marketing, finance, management decision-making information systems (Aral, 2011; Goh *et al.*, 2013), and product feature identification (Archak *et al.*, 2011). Box office revenue will be diluted and will decrease due to the effect of WOM impact (Yang *et al.*, 2012). Quantitative aspects of consumer satisfaction rate or review rate have been used for predicting sales and purchase of goods such as books (Chevalier & Mayzlin, 2006) and movies (Asur & Huberman, 2010; Duan *et al.*, 2008b; Gaikar *et al.*, 2015), and repeat purchase of goods such as beauty products (Clemons *et al.*, 2006), luxury products (Liu *et al.*, 2012a), and fragrance (Moe *et al.*, 2011), for estimating aggregate economic outcomes with UGC.

It is more important to examine different characters than single attributes of WOM. Box office revenue forecasting could be effectively done by combining an additional PN ratio (ratio of positive to negative tweets) with the base model. The combination of UGC and impact of negative sentiment with base model was revealed its box office revenue (Asur & Huberman, 2010). Further, the percentage of negative and positive opinion has been examined, and it was identified that consumers recognized WOM with high value. It is a better sales forecasting variable for electronic product sales forecasting (Chong *et al.*, 2016b).

The characteristics of the online opinions such as volume, sentiment value of UGC, and marketing characteristics such as promotional strategies (discount rate, free delivery) with interplay effects have been used to predict product sales. The interplay of variables has a significant improvement on sales prediction (Chong *et al.*, 2016). Therefore, the combination of UGC has become a very important predictive mechanism for sales forecasting.

Most studies forecasted sales for the automobile, fashion, movie, book, electronic product, and service industries. Innovative e-commerce retail products have poor product shelf-life cycles and the demand for such products is changing within a day or even hours (Kolchyna, 2017).

According to literature findings, UGC has a stronger predictive capability, and negative conclusions of recommendations are impactful and influential for product future performances. If the company can be flexible in hard and soft infrastructure, it can provide better consumer satisfaction and product improvements (Subramanian *et al.*, 2016). Therefore, short-duration flexibility and forecasting, such as following a daily or weekly forecast, are important for innovative products to get opportunities in a competitive market. Further, daily forecasting is rare for innovative products including UGC. There is space to fill the gap by using consumer text-based opinions (Balachandra *et al.*, 2020; Hou *et al.*, 2017; Sugathadasa *et al.*, 2021; Zhu & Zhang, 2010). Therefore, "Impact of UGC on innovative product sales performances" is established as a research question.

The research objectives are stated as follows: first, to identify the factors that impact consumer behavior for innovative product purchasing; second, to develop a demand forecasting model for innovative FMCG product sales; and third, to identify the importance of UGC to improve the demand forecasting accuracy of innovative products.

3. Methodology

3.1. *Research design*

Data were collected from an online innovative product selling shop on which users are active and which considered the users' views to function. We mainly focused on the innovative product portfolio based on daily FMCG sales and UGC according to respective products (Kakatkar *et al.*, 2020). The end user's information was analyzed to identify the potential requirement of new users. Data consisted of 2502 consumers' opinions from 2019/01/02 to 2019/05/27. The sentiment index extraction analytical technique was used to identify each consumer opinion to develop the forecasting model. The sentiment value was generated using the VADER application. The UGC was selected which was important for innovative FMCG product sales forecasting. The variables were selected according to literature findings (Table 1). Those variables used as predictive variables to develop forecasting models. The prophet forecasting model was deployed to forecast the next 10 days' sales using R software. Specific

Table 1.   Variable summary.

Variable	Description	Operationalization	Citation
Average positive sentiment strength (APSV)	The average positive sentiment value extracted from online opinions.	The positive sentiment of each positive opinion was extracted, and then the average value from 0 to 1 was taken.	Chong *et al.*, 2016b; Hou *et al.*, 2017; Kasturia *et al.*, 2020
Average negative sentiment strength (ANSV)	The average negative sentiment value extracted from online opinions.	The negative sentiment of each negative opinion was extracted, and then the average value from 0 to −1 was taken.	Chong *et al.*, 2016b; Hou *et al.*, 2017
Positive opinion volume (TNPR)	The number of all positive online opinions.	The number of all positive opinions.	Fan *et al.*, 2017
Negative opinion volume (TNNR)	The number of all negative online opinions.	The overall number of negative opinions.	Fan *et al.*, 2017
Percentage of negative opinions (NP)	The proportion of negative opinions in total opinions count.	The percentage of negative opinions, computed by dividing the total number of negative opinions by the total number of opinions.	Kolchyna, 2017; Chong *et al.*, 2016b
Percentage of positive opinions (PP)	The proportion of positive opinions in total opinions count.	The percentage of positive opinions, computed by dividing the total number of positive opinions by the total volume of opinions.	Kolchyna, 2017; Chong *et al.*, 2016b

measures MAPE, R^2, and RMSE were used to check the accuracy of the model.

The innovative product portfolio consisted of personal care products because those products are highly based on others' recommendations (Zunic & Donko, 2020). Sentiment analysis was used to classify the unstructured text data to develop a regressor. The entire process of sentiment analysis will be explained in Section 3.3.

3.2. *Sentiment analysis*

A consumer can explain their opinion, judgment, attitude, and emotion toward products or services in various ways. The process of computational identification and classification of textual opinions, in particular to determine whether a consumer's attitude towards a particular topic, product, etc. is positive, negative or neutral is known as sentiments analysis (Liu *et al.*, 2012b). Consumer opinion can be divided into two parts, objective and subjective. Opinions that are objective do not express a viewpoint or use emotional language, but are based on factual information. Opinions that represent the user's viewpoints are classified as subjective (Duan *et al.*, 2008a). The bag of word analysis methods is also used in text classification, but this chapter mainly focuses on daily consumer opinion; the bag of words classifying approach is not suitable for this type of brief opinion sentiment analysis. Because those opinions are very brief in content, compared to the many words in the word documents, the bag of words classifier will not give high accuracy and precision for this type of social media analytics (Cui *et al.*, 2017; Wang *et al.*, 2012). Subjective opinion was selected, and the lexicon-based approach was used to analyze sentiment value from UGC. The VADER dictionary-based analytical approach was used to calculate the intensity of UGC and polarity classification. The application is provided by the Natural Language Tool Kits package (NLTK).

3.3. *Opinion selection and sentiment analysis*

The impact of opinion was considered as a time function. When the opinion gets old, it becomes less influential for another consumer. Since readers tend to use more recent information, old information is neglected. Therefore, the impact of the opinion diminishes over time. It has a downtrend impact on the sales, and more recent opinions impact sales more

(Chern *et al.*, 2014). Newly generated consumer opinions' sentiment value was used for sales forecasting (Kolchyna, 2017). The strength of positive, negative, and neutral sentiment values was measured. Each UGC is considered a unit of analysis (Hou *et al.*, 2017). Each word of the consumer's opinion was checked with the VADER lexicon to identify the valence and relevant strength of the given text. The intensity of the opinion was calculated between +1 and −1 representing strong positive and strong negative, respectively. The neutral opinion was assigned 0 strength (Kasturia *et al.*, 2020). Finally, the average sentiment value was analyzed each day.

3.4. *Variables summary*

Table 1 represents the variables that were used as extra regressors as WOM characteristics to improve the forecasting accuracy.

3.5. *The prophet forecasting model*

Facebook's core data science team has published the prophet forecasting model and it can be used to identify nonlinear trends that fit with time such as daily, weekly, seasonality, and holidays. The prophet time series forecasting model is a decomposable time series model and it has three main components: trends, seasonality, and holidays.

$$Y(t) = g(t) + s(t) + h(t) + \varepsilon(t) \tag{1}$$

$g(t)$ — piecewise linear or logistic growth curve for modeling non-periodic changes in time series,
$s(t)$ — periodic changes (e.g., weekly, yearly, seasonality),
$h(t)$ — effects of holidays (user-provided) with irregular schedules,
$\varepsilon(t)$ — errors term accounts for any unusual changes not accommodated by the model.

The model is a generalized additive model (GAM). It will quickly and easily fit with newly added data. A special data frame has been provided by the prophet model. It contains two main columns representing time as "*ds*", and another column represents target variable (dependent variable) "*y*", and it stores time series data in the data frame. In the development of *ds, y* is the only pre-processing step that needs to be done in the prophet model. Due to the special data framework of the model, it can be used to improve the forecasting accuracy by adding extra regressors. For example, first, the

Table 2.   Model comparison.

Model	R^2 value	RMSE	MAPE (%)	P-value	Reduction of MAPE (%)
Base prophet model	0.3058	0.56106	16.97	2.28E-11	
Base model with sentiment value	0.347	0.52693	15.75	4.96E-13	7
Model with volume of opinion	**0.7678**	**0.28263**	**8.51**	**2.20E-16**	**50**
Model with volume * sentiment value	**0.7046**	**0.34535**	**9.92**	**2.20E-16**	**42**
Model with positive percentage	0.6336	0.573943	18.13	2.20E-16	−7
Model with negative percentage	0.6336	0.5739	18.13	2.20E-16	−7

basic model was developed by using daily sales volume to forecast the next 10 days' sales. The accuracy of the model was checked with the test dataset (see Table 2). The base model's accuracy was low. Second, UGC sentiment impact was forecasted for the next 10 days as a separate function of positive impact and negative impact of UGC. Third, the sentiment value forecasted data were added to the base forecasting model as an extra regressor to predict next 10 days' sales. Fourth, the new model's accuracy (Base model with sentiment value) was tested and the model's improvement was compared relative to base model. The base model's accuracy was developed by using selected variables as five different models. Finally, variables which improved the model's accuracy were selected for user-sensitive FMCG sales forecasting.

4. Findings

Five different methods were developed using UGC. After the data cleaning process, the sentiment value of WOM was generated. The basic forecasting model improved with separate functions of each variable (positive and negative) because NWOM has a completely different impact than PWOM on sales performances (Iftikhar & Khan, 2020).

4.1. *Model comparison with performance matrix*

The daily sales volume has been transformed as log values, and the relative percentage improvement of each model is calculated. The base prophet model is considered the benchmark to evaluate the each model (Kolchyna, 2017). The better performing models and the models showing relative improvement are in bold in Table 2.

The base model's accuracy can be improved by using WOM attributes as an extra regressor. The base forecasting model shows low performance values because some predicted valued have large errors. These may not be applicable to sales predictions (see Figure 1 and Table 2). The first model improved upon using sentiment value and it has some relative improvement when it is used as a regressor. The volume extra regressor is used for the second model with a separate function of positive and negative volumes. The volume of the UGC (Positive and Negative) shows better performance than the base model and other developed models. When there is a high volume of opinion, the impact may be low due to their emotional opinions on the e-commerce platform and vice-versa. Therefore, the third variable is considered an interplay combination with volume and sentiment. It performed well compared to the base model and sentiment model.

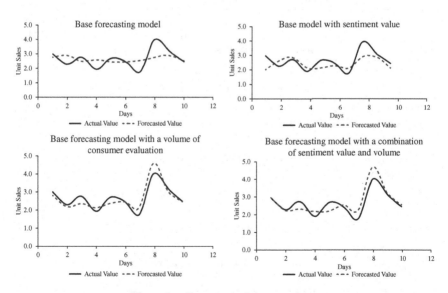

Figure 1.   Forecast model comparison.

The fourth model is developed by using positive and negative percentages as a separate function. But, models with positive and negative percentages were not performing well in relation to innovative FMCG sales (see Table 2-statistics). Therefore, the forecasting model's accuracy can be improved with the use of WOM characteristics in e-commerce.

4.2. *Variable improvement*

The combination of the above characteristics performed better than those variables used as an individual regressor. All UGC variables can be used to improve unit sales forecasting accuracy at a 95% significant interval.

Table 3 and Figure 2 illustrate the relative improvement of base forecasting model when the interplay variable is used as an extra regressor. The relative improvement of the MAPE value of the model is shown in bold.

The sentiment of the user opinion may impact the sales performance even when there is a low volume percentage. Therefore, the fourth model was improved with a combination of sentiments. The impact of text

Table 3.   Improvement of variable summery.

Model	R^2 value	RMSE	MAPE (%)	*P*-value	Reduction of MAPE (%)
Base prophet model	0.3058	0.56106	16.97	2.28E-11	
Model with negative sentiment value	0.347	0.5363	17.35	4.96E-13	–2
Model with negative percentage	0.6336	0.5739	18.13	2.20E-16	–7
Model with NP * ANSV	**0.5686**	**0.5041**	**15.23**	**2.20E-16**	**10**
Model with positive sentiment value	0.3319	0.5340	17.08	2.07E-12	–1
Model with positive percentage	0.6336	0.5739	18.13	2.20E-16	–7
Model with PP * APSV	**0.5093**	**0.4586**	**14.73**	**2.20E-16**	**13**

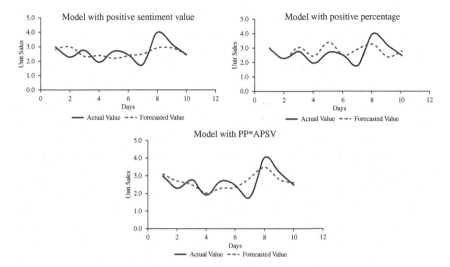

Figure 2.   Model comparison.

opinion on sales performance is high and those models were improved when considering sentiments. Therefore, it is better to consider interplay variables in sales prediction by using NWOM characteristics.

5. Discussion of Results

This research intended to identify the impact of UGC on user-sensitive products' sales forecasting in an innovation-driven market. The UGC attributes were segregated as predictive variables to improve sales fore-casting. Finally, we found the most significant variables for user-sensitive products' sales forecasting.

This chapter will point you in the right direction to improve five ana-lytical procedures that have been identified as critical in order to improv-ing the technique to forecast innovation.

First, sentiment (text-based) opinion had a very powerful predictive capability on sales. Sentiment based on positive and negative sentiment values was used to perform a forecasting model and it did not lead to a proper improvement over the base prophet model. The relative MAPE reduction was 7%. But, many researchers have concluded that sentiment has a significant impact on innovative products as well as innovative man-agers' decisions. According to Chong *et al.* (2016b), sentiment has less impact based on the search product category. It could conclude that the sentiment of UGC has low predictive power when it was used as an individual regressor, relating to user-sensitive innovative products.

Second, the base model was developed with consumer-generated volume, and it makes a more accurate forecasting model when compared with other selected extra regressors. When there are more user opinions per day or newly generated user opinions, it allows one to change the direction of the sales and it will help increase the predictive accuracy. Because it implies that the product is popular among customers, and it can raise awareness and attract attention to the product, UGC volume is considered to have an impact on sales. The relative MAPE value reduction of the base model was 50%. Many innovative analyses suggested that UGC opinion's volume could be used as a predictive variable for sales forecasting based on different product categories. It could be justified according to the user-sensitive products. The analytics, which includes a risk assessment and a few alternative scenarios, predicts what will happen in the future with a reasonable level of confidence. Predictive innovative analytics is a business tool that analyzes current data and past facts to better understand customers, goods, and partners, as well as identify potential consequences and possibilities.

Third, research has confirmed that all selected variables make a significant contribution to improve sales forecasting. Although both volume and sentiment had predictive power, the volume represented a more significant capability on unit sales than sentiment. But, sentiment contributes more significantly, and it effectively improved the model's accuracy by integrating with other variables. For instance, the combination of sentiment and volume had better predictive capability than using sentiment alone. The relative MAPE reduction of the combination variable was 42%. But, it does not exceed the performance of volume as compared to volume used separately. It can be concluded that when there are more user opinions, the combination of volume and sentiment has better capability on sales forecasting than using sentiment alone.

Fourth, Hou *et al.* (2017) suggested that negative valance and negative opinion's volume had a significant impact on sales. When there is a negative opinion of a product or brand, it is kept in mind when someone purchases the product. Therefore, it was used as a separate function to improve the accuracy. According to Chong *et al.* (2016b), it justified that negative and positive percentage performances play a vital part as a predictor in forecasting; here, it did not perform well compared to the base model, but it could improve with a combination of sentiment in both positive and negative volume percentages. Therefore, it is better to consider both positive opinion and negative opinion to understand consumer behavior. Since consumers especially pay attention to more polarized and text-based opinions, the user interaction could be effectively identified by using these combinations than using them individually.

Finally, it can be concluded that, though sentiment makes a less significant improvement with the prophet model, it could improve with a combination of other attributes. The interplays between sentiment and other attributes make a more important contribution to product sales. The volume had been studied as a percentage value (Chong *et al.*, 2016b). That variable can be further improved with a combination of sentiments and used to improve forecasting accuracy. According to the background study, this variable was not investigated based on online UGC. Consumer-generated volume, sentiment, and percentage values in both aspects will impact consumer purchasing behavior in an innovation-driven market environment.

According to the findings of the research, business practitioners should concentrate their marketing efforts on e-platform techniques to enhance sales, such as increasing online visibility (by increasing the volume and valence of e-UGC) and offering a venue for potential customers to directly inquire about vendors' products; facilitating a platform that allows users to obtain knowledge about products based on user experiences will bring value to e-platform sales and predictions.

6. Conclusion and Implication

The trend of digital media use for product purchasing, marketing have increased and it has rapid growth in innovative business environment. Then, eWOM plays a massive role in contributing to change the direction of innovative product performances. This chapter focused on identifying how UGC impacts user-sensitive FMCG products. The objective was to identify the factors that cause a change in consumer purchasing behavior. An FMCG product sold by an online platform was selected to collect sales information, and consumer product opinions were extracted. Consumer opinion was analyzed by using the VADER sentiment analysis, and forecasting models were developed by using the prophet forecasting model. Finally, suitable variables were selected that showed better performances based on lower MAPE and RMSE values.

This analysis showed that all selected variables have a significant impact on and influential factors with regard to user-sensitive product sales. Many of the studies have demonstrated that consumer opinion volume and valence play an important role in sales prediction. Previous studies gradually shifted focus to understand the text-based opinion of users from numeric opinion. The study of Chong *et al.* (2016b) demonstrated that the combination of UGC plays a more important role than using a single variable to predict sales. According to the literature survey, the selected attributes were applied to the FMCG product perspective and demonstrated that this combination is very applicable to the user-sensitive FMCG product perspective.

Unique findings that relate to eWOM could be demonstrated from the study. According to Chong *et al.* (2016b) and Kolchyna (2017), it has been demonstrated that the percentage of user opinions (both positive and negative) had a better predictive capability, but here, the accuracy of the model was low. At present, users tend to read others' opinions rather than paying attention to the volume of the opinions. Even though the volume percentage had predicting capability, it may change due to sentiment opinions. Therefore, the combination of sentiment and percentage (both positive and negative) performs well according to the FMCG product category; this is considered an extension of previous studies (Kolchyna, 2017; Chong *et al.*, 2016b).

The identified innovative findings provide better guidance for e-commerce management as well as marketing strategies to optimize the use of resources. This method aids in predicting the impact of future decisions so that adjustments can be made before it is too late. This greatly enhances decision-making because future consequences are factored into the prediction. One can improve the optimization of scheduling, manufacturing, supply chain design, and inventory to provide customers what they want in the most efficient manner. UGC-based interaction is very important to sales. Then, paying attention to the consumer opinions and by providing a platform to generate more opinions leads to understand sales performances. Real-time consumer opinions gained by text mining lead to better supply chain implementation from strategic purchasing. Each year, the demand for real-time data grows, and reliable updates are essential for building effective strategies for dealing with difficult situations. With additional variables in the mix, forecasting will undoubtedly become much more important in formulating correct corporate reactions and strategies for future initiatives. Therefore, one should not only consider the number of users but also the sentiment of the user and other presented factors for predicting sales. Furthermore, integrating live dashboards will enable businesses to quickly obtain pertinent information about their operations and respond if any possible difficulties occur. Therefore, the proposed innovative analytical approach provides a new methodology to present theories that relate to e-commerce business environment, decision-making, as well as managing resources.

6.1. *Limitation and future direction*

This research mainly considered the personal care and home care product portfolio since purchasing such products is based on others' recommendations. Future research should be based on SKU (Stock Keeping Units), and one should also investigate whether the features and approaches of this study apply to other SKUs. Moreover, different types of products can be used for further improvement of sales prediction (quick perishable items such as vegetables and fruits distribution). The model's accuracy can be compared with other time series forecasting methods to check the importance of regressors in forecasting.

This chapter mainly focused on text-based analysis. Future research should investigate the impact of different UGC approaches such as picture recognition or video recognition on social media. Further, different types of social media platforms are maintained by companies. These platforms can be used to compare and provide a greater understanding of consumer behavior across various platforms. The companies can find out which media-based behavior will impact product performance changes massively and can improve their marketing strategies based on the selected media. It may help to save resources and launch new innovative strategies to attract more users with a competitive market advantage. Further, online-based companies are providing a platform to provide user opinions. Since there are negative opinions, it may impact product sales. The control of negative opinions is essential for product performance. According to Chong *et al.* (2016b), answering consumer questions is a good predictive component for sales forecasting. Therefore, the new approach should investigate whether the impact of negative consumer opinions can be minimized by sentiment-based answering of consumer questions and opinions. This will lead to innovative answering approaches.

Acknowledgment

The authors would like to acknowledge the financial support given by the University of Moratuwa, Sri Lanka, through the University Senate Research Committee research grant: SRC/LT/2021/22 for this research.

References

Andreassen, T. W. & Streukens, S. (2009). Service innovation and electronic word of mouth: Is it worth listening to? *Managing Service Quality: An International Journal,* 19(3), 249–265. https://doi.org/10.1108/09604520 910955294.

Antweiler, W. & Frank, M. Z. (2004). Is all that talk just noise? The information content of internet stock message boards. *Journal of Finance,* 59(3), 1259–1294. https://doi.org/10.1111/j.1540-6261.2004.00662.x.

Aral, S. (2011). Commentary — Identifying social influence: A comment on opinion leadership and social contagion in new product diffusion. *Marketing Science,* 30(2), 217–223. https://doi.org/10.1287/mksc.1100.0596.

Archak, N., Anindya, G., & Ipeirotis, P., G. (2011). Deriving the pricing power of product features by mining consumer reviews. *Management Science,* 57(8), 1485–1509. https://doi.org/10.1287/mnsc.1110.1370.

Asur, S. & Huberman, B. A. (2010). Predicting the future with social media. *Proceedings — 2010 IEEE/WIC/ACM International Conference on Web Intelligence, WI 2010,* 1, 492–499. https://doi.org/10.1109/WI-IAT. 2010.63.

Balachandra, K., Perera, H. N., & Thibbotuwawa, A. (2020). Human factor in forecasting and behavioral inventory decisions: A system dynamics perspective. In *International Conference on Dynamics in Logistics* (pp. 516–526). Cham: Springer.

Campbell, C., Ferraro, C., & Sands, S. (2014). Segmenting consumer reactions to social network marketing. *European Journal of Marketing,* 48(3–4), 432–452. https://doi.org/10.1108/EJM-03-2012-0165.

Chen, Y. & Xie, J. (2008). Online consumer review: Word-of-mouth as a new element of marketing communication mix. *Management Science,* 54(3), 477–491. https://doi.org/10.1287/mnsc.1070.0810.

Chern, C., Wei, C., Shen, F., & Fan, Y. (2014). A sales forecasting model for consumer products based on the influence of online word-of-mouth. *Information Systems and E-Business Management,* 13(3), 445–473.

Cheung, C. M. K. & Thadani, D. R. (2012). The impact of electronic word-of-mouth communication: A literature analysis and integrative model. *Decision Support Systems,* 54(1), 461–470. https://doi.org/10.1016/j.dss.2012.06.008.

Chevalier, J. A. & Mayzlin, D. (2006). The effect of word of mouth on sales: Online book reviews. *Journal of Marketing Research*, 43(3), 345–354. https://doi.org/10.1509/jmkr.43.3.345.

Chong, A. Y. L., Li, B., Ngai, E. W. T., Ch'ng, E., & Lee, F. (2016a). Predicting online product sales via online reviews, sentiments, and promotion strategies. *International Journal of Operations and Production Management*, 36(4), 358–383. https://doi.org/10.1108/IJOPM-03-2015-0151.

Clemons, E. K., Gao, G., & Hitt, L. M. (2006). When online reviews meet hyper-differentiation: A study of the craft beer industry. *Journal of Management Information Systems*, 23(2), 149–171. https://doi.org/10.2753/mis0742-1222230207.

Cui, R., Moreno, A., & Zhang, D. J. (2017). The operational value of social media information. *Production and Operations Management*, 27(10), 1749–1769. https://doi.org/10.1111/poms.12707.

Das, S. R. & Chen, M. Y. (2007). Yahoo! for Amazon: Sentiment extraction from small talk on the web. *Management Science*, 53(9), 1375–1388. https://doi.org/10.1287/mnsc.1070.0704.

Deloitte. (2015). Future of E-commerce: Uncovering innovation. pp. 23–26. https://www2.deloitte.com/content/dam/Deloitte/in/Documents/technology-media-telecommunications/in-tmt-future-of-e-commerce-noexp.pdf.

Duan, W., Gu, B., & Whinston, A. B. (2008a). Do online reviews matter? — An empirical investigation of panel data. *Decision Support Systems*, 45(4), 1007–1016. https://doi.org/10.1016/j.dss.2008.04.001.

Duan, W., Gu, B., & Whinston, A. B. (2008b). The dynamics of online word-of-mouth and product sales — An empirical investigation of the movie industry. *Journal of Retailing*, 84(2), 233–242. https://doi.org/10.1016/j.jretai.2008.04.005.

East, R., Hammond, K., & Lomax, W. (2008). Measuring the impact of positive and negative word of mouth on brand purchase probability. *International Journal of Research in Marketing*, 25(3), 215–224. https://doi.org/10.1016/j.ijresmar.2008.04.001.

Fan, Z. P., Che, Y. J., & Chen, Z. Y. (2017). Product sales forecasting using online reviews and historical sales data: A method combining the bass model and sentiment analysis. *Journal of Business Research*, 74, 90–100. https://doi.org/10.1016/j.jbusres.2017.01.010.

Fang, X. & Zhan, J. (2015). Sentiment analysis using product review data. *Journal of Big Data*, 2(1). https://doi.org/10.1186/s40537-015-0015-2.

Gaikar, D. D., Marakarkandy, B., & Dasgupta, C. (2015). Using Twitter data to predict the performance of bollywood movies. *Industrial Management and Data Systems*, 115(9), 1604–1621. https://doi.org/10.1108/IMDS-04-2015-0145.

Goh, K. Y., Heng, C. S., & Lin, Z. (2013). Social media brand community and consumer behavior: Quantifying the relative impact of user- and marketer-generated content. *Information Systems Research*, 24(1), 88–107. https://doi.org/10.1287/isre.1120.0469.

Grützmann, A., Zambalde, A. L., & de Souza Bermejo, P. H. (2019). Inovação, Desenvolvimento de Novos Produtos e as Tecnologias Internet: Estudo Em Empresas Brasileiras. *Gestão & Produção*, 26(1), 1–15. https://doi.org/10.1590/0104-530x1451-19.

Guo, Y., Fan, D., & Zhang, X. (2020). Social media–based customer service and firm reputation. *International Journal of Operations and Production Management*, 40(5), 575–601. https://doi.org/10.1108/IJOPM-04-2019-0315.

Hennig-Thurau, T., Gwinner, K. P., Walsh, G., & Gremler, D. D. (2004). Electronic word-of-mouth via consumer-opinion platforms: What motivates consumers to articulate themselves on the internet? *Journal of Interactive Marketing*, 18(1), 38–52. https://doi.org/10.1002/dir.10073.

Hou, F., Li, B., Chong, A. Y., Yannopoulou, N., & Liu, M. J. (2017). The management of operations understanding and predicting what influence online product sales? A neural network approach. *Production Planning & Control*, 7287(July), 1–12. https://doi.org/10.1080/09537287.2017.1336791.

Iftikhar, R. & Saud Khan, M. (2020). Social media Big Data analytics for demand forecasting. *Journal of Global Information Management*, 28(1), 103–120. https://doi.org/10.4018/JGIM.2020010106.

Kakatkar, C., Bilgram, V., & Füller, J. (2020). Innovation analytics: Leveraging artificial intelligence in the innovation process. *Business Horizons*, 63(2), 171–181. https://doi.org/10.1016/j.bushor.2019.10.006.

Kasturia, V., Sharma, S., & Sharma, S. (2020). Automatic product saleability prediction using sentiment analysis on user reviews. In *Proceedings of the Confluence 2020 — 10th International Conference on Cloud Computing, Data Science and Engineering*.

Keaveney, S. M. (1995). Customer switching behavior in service industries: An exploratory study. *Journal of Marketing*, 59(2), 71. https://doi.org/10.2307/1252074.

Kolchyna, O. (2017). Evaluating the impact of social-media on sales forecasting: A quantitative study of world biggest brands using Twitter, Facebook and Google Trends.

Liu, B. F., Jin, Y., Briones, R., & Kuch, B. (2012a). Managing turbulence in the blogosphere: Evaluating the blog-mediated crisis communication model with the American red cross. *Journal of Public Relations Research*, 24(4), 353–370. https://doi.org/10.1080/1062726X.2012.689901.

Liu, F., Li, J., Mizerski, D., & Soh, H. (2012b). Self-congruity, brand attitude, and brand loyalty: A study on luxury brands. *European Journal of Marketing*, 46(7), 922–937. https://doi.org/10.1108/03090561211230098.

Liu, Y., Huang, X., An, A., & Yu, X. (2007). ARSA: A sentiment-aware model for predicting sales performance using blogs. In *Proceedings of the 30th Annual International ACM SIGIR Conference on Research and Development in Information Retrieval, SIGIR'07*.

Mayzlin, D. (2006). Promotional chat on the Internet. *Marketing Science*, 25(2), 155–163. https://doi.org/10.1287/mksc.1050.0137.

Moe, W. W., Trusov, M., & Smith, R. H. (2011). The value of social dynamics in online product ratings forums. *Journal of Marketing Research*, 48(3), 444–456. https://doi.org/10.1509/jmkr.48.3.444.

Ramanathan, U., Subramanian, N., & Parrott, G. (2017). Customer satisfaction role of social media in retail network operations and marketing to enhance customer satisfaction. *International Journal of Operations & Production Management*, 37(1), 105–123. https://doi.org/10.1108/IJOPM-03-2015-0153.

Rozin, P. & Royzman, E. B. (2001). Negativity bias, negativity dominance, and contagion. *Personality and Social Psychology Review*, 5(4), 296–320. https://doi.org/10.1207/S15327957PSPR0504_2.

Schaer, O., Kourentzes, N., & Fildes, R. (2018). Demand forecasting with user-generated online information. *International Journal of Forecasting*, 35(1), 197–212. https://doi.org/10.1016/j.ijforecast.2018.03.005.

Singh, A., Shukla, N., & Mishra, N. (2018). Social media data analytics to improve supply chain management in food industries. *Transportation Research Part E: Logistics and Transportation Review*, 114, 398–415. https://doi.org/10.1016/j.tre.2017.05.008.

Subramanian, N., Gunasekaran, A., & Gao, Y. (2016). Innovative service satisfaction and customer promotion behavior in the Chinese budget hotel: An empirical study. *International Journal of Production Economics*, 171, 201–210. https://doi.org/10.1016/j.ijpe.2015.09.025.

Sugathadasa, R., Wakkumbura, D., Perera, H. N., & Thibbotuwawa, A. (2021). Analysis of risk factors for temperature-controlled warehouses. *Operations and Supply Chain Management: An International Journal*, 14(3), 320–337.

Themistocleous, M., Irani, Z., & Love, P. E. D. (2004). Evaluating the integration of supply chain information systems: A case study. *European Journal of Operational Research*, 159(2), 393–405. https://doi.org/10.1016/j.ejor.2003.08.023.

Tirunillai, S. & Tellis, G. J. (2012). Does chatter really matter? Dynamics of user-generated content and stock performance. *Marketing Science*, 31(2), 198–215. https://doi.org/10.1287/mksc.1110.0682.

Wang, H., Can, D., Kazemzadeh, A., Bar, F., & Narayanan, S. (2012). A system for real-time Twitter sentiment analysis of 2012 U.S. presidential election cycle. In *Proceedings of the 50th Annual Meeting of the Association for Computational Linguistics*. https://doi.org/10.1145/1935826.1935854.

Wu, L., Lou, B., & Hitt, L. (2019). Data analytics supports decentralized innovation. *Management Science.* https://doi.org/10.1287/mnsc.2019.3344.

Yang, J., Kim, W., Amblee, N., & Jeong, J. (2012). The heterogeneous effect of WOM on product sales: Why the effect of WOM valence is mixed? *European Journal of Marketing*, 46(11), 1523–1538. https://doi.org/10.1108/03090561211259961.

Yu, J., Subramanian, N., Ning, K., & Edwards, D. (2015). Product delivery service provider selection and customer satisfaction in the era of Internet of Things: A Chinese e-retailers' perspective. *International Journal of Production Economics*, 159, 104–116. https://doi.org/10.1016/j.ijpe.2014.09.031.

Zhang, C., Tian, Y. X., Fan, Z. P., Liu, Y., & Fan, L. W. (2020). Product Sales forecasting using macroeconomic indicators and online reviews: A method combining prospect theory and sentiment analysis. *Soft Computing*, 24(9), 6213–6226. https://doi.org/10.1007/s00500-018-03742-1.

Zhu, F. & Zhang, X. (2010). Impact of online consumer reviews on sales: The moderating role of product and consumer characteristics. *Journal of Marketing*, 74(2), 133–148. https://doi.org/10.1509/jmkg.74.2.133.

Zunic, E. & Donko, D. (2020). Algorithm for successful sales forecasting based on real — World data. (April). https://doi.org/10.5121/ijcsit.2020.12203.

Chapter 13

Conclusion

Nachiappan Subramanian[*,§], **S. G. Ponnambalam**[†,¶]
and Mukund Janardhanan[‡,‖]

University of Sussex Business School, Brighton, UK
†*Vellore Institute of Technology, Vellore, India*
‡*School of Engineering, University of Leicester,*
Leicester, UK
§*n.subramanian@sussex.ac.uk*
¶*ponnambalam.g@vit.ac.in*
‖*mukund.janardhanan@leicester.ac.uk*

Overall, the aim of this book is to explain nascent topics within innovation analytics and suggest future research directions and meaningful use of data engineering with advanced AI-based techniques that could contribute to radical and incremental product and process innovations. Innovation is a function of performance and time. A higher probability for success of innovation depends on continuous improvements and reduction in time. Based on the discussion and insights offered under the three themes in this book, this chapter summarizes the potential research pathways in the innovation analytics domain. These are only indicative and there are plenty of other research gaps to consider.

1. Product and Process Innovation

Chapters 2–5 explained the possibilities of integration of data engineering, AI, or emerging analytics for product or process innovation. Interestingly, the authors have focused on innovation from the customer perspectives, manufacturing, R&D, and business model perspectives. The common theme emerging from all the studies informed the need for an interface to understand the silo effects, i.e., from a data point of view, analytics point of view, and translation and application of those to product and process innovation. For the benefit of the readers, this section summarizes the major research gaps spelled out by the respective authors.

1.1. *Customer perspective*

In terms of understanding customer capabilities, in the future, a designer can engage with emerging data and analytics to consider the concept of affordance when applying analytics for innovation. In addition, analytics could heavily support future researchers to disentangle the softer aspects of human and machine collaboration. The authors in the book have emphasized the need to integrate the information available in silo data sources, efforts to micro-match timescales between product development and rapidly changing consumer trends and need for integrated e-commerce and web/social media insights mining capabilities. In particular, there is immense scope for natural language processing and third-wave AI (contextual adaptation, few data for training, conversation with minimal samples of training) to handle vast consumer data to innovate in a short span of time.

1.2. *Manufacturing perspective*

To scale up the innovations in smart manufacturing implementation, there is a great need for development of AI-based techniques to consider multiple factors. The use of big data and IoT needs to be investigated from shop floor to intercompany level to enhance process-level incremental and radical simulations. The list of AI applications to drive innovation is potentially endless in manufacturing, and the current capabilities are improving all the time (Arinez *et al.*, 2020). Shop floor engineers can use the discoveries of AI and learn new information on shop floor problems, which leads to new innovative solutions. Moreover, the engineers can remake experiences in real time at all levels of the shop floor. Such innovations can have a significant impact on the bottom line of a company and its growth trajectory.

1.3. *R&D perspective*

There is immense scope for cloud-enabled platforms and advanced analytics to support next-generation R&D. The need for open and interoperable systems has also been reported by various practitioners to offer advanced analytics through platforms as service approaches. In this book, the authors have argued on the need for a system architecture approach for designing multidisciplinary products. It is well known that many companies are using only 5% of the data available in their R&D activities because of shortage of data analysts, collaborative work with the team, etc. In addition, the pharmaceutical sector R&D, specifically, has extensively used data analytics for new drug development and to personalize drugs compared to other sectors. So, there is a need for big data and AI application for early customization in other sectors that will enhance product and process innovations.

1.4. *Business model perspective*

A recent study reports the lack of research investigating the relationship between business model design and its impact on R&D performance, especially the role of learning capabilities through the support of big data and the empirical evidence (Sun & Liu, 2021). The research also suggests that one should analyze the interaction of big data analytics talent capability, ambidextrous learning capabilities, cloud ERP, and cultural factors between business model design and process & product innovation.

2. Artificial Intelligence

The evolution of AI with the support of algorithmic improvement and hardware development is supporting innovation, but there is still a long way to go. Experts suggest that one should develop unsupervised learning compared to supervised learning which will certainly limit the human intervention to the maximum possible extent and not exactly mimic nature in the development of new algorithms (IBM, 2021). This book discussed the connection between big data and AI, compared AIs and humans, and analyzed AI application for reducing food waste and traffic light management. A few suggestions offered by the authors are as follows.

There is a need for a great deal of research in understanding the integration of AI in the industrial integration of IoT and evolving technological challenges such as cyber-physical security, real-time analytics, and edge computing. In addition, there is a big question regarding trust and the unintended use of technology along with AI. A large-scale study is essential to determine the unethical use of AI and its consequences and how to prevent misuse in the future. On the other hand, from the application point of view, use of images and videos as an input data to AI is a potential future research pathway. Similarly, analyzing the technical and business impact of integration of AI with distributed ledger technology is an obvious way forward for researchers to develop insights. On the supply chain innovation front, AI has a significant role to play in monitoring the flow of products from source to destination and to support the decision-makers in dealing with uncertain complex challenges.

In the decision-making context, AI is in early days of development in cognitive reasoning and deep reasoning at match human level reasoning. It is predicted that, in the future, the emergence of data, deep-learning algorithms, and graphical processing units will enable one to scale up reasoning computations (IBM, 2021). As per Paschen *et al.* (2020), AI has a substantial role to play in competence-enhancing product (customer order optimization) and process (predictive maintenance) innovation rather than competence-destroying product (autonomous trucks/cars) and process innovation (automated bids). Furthermore, there is a need for further research to (i) develop standards for seamless application programming interfaces of AI with different software applications, (ii) ensure training datasets without human biases and cultural assumptions, and (iii) carefully design privacy measures for AI integration.

3. Data Engineering

Data-driven innovation has benefited leading tech firms. Social media is extensively used in marketing, but there are prospective directions to those in product and process innovations. There are serious efforts by technological companies to gather longitudinal data about the customer that can support innovation. Also, collection of user data through social media under the experimental setting is considered to be sensible in the current context. All the chapters included in the book have suggested that one should develop an accurate model for forecasting and support of operational decisions with further discussion on capturing data for creating and generating ideas instead of routine development. Furthermore, Luo (2022) suggested that data-driven innovation required one to mine data from multiple sources such as e-commerce sites, social media, open-source repositories, patent and publication databases, and archival data in addition to digital foot print data from failed experiments available in the in-house repository and open innovation contexts. In particular, there is an ongoing debate on the use of predictive analytics for innovation where we look back on past data and trends to predict the future. However, the recent availability of advanced data through multiple channels will enable companies to use analytics effectively to derive newer ideas. This is possible with the right combination of predictive analytics and innovation. A recent study by Barsoux (2022) articulated well the critical role of external data or outside perspectives from non-traditional sources contributions in achieving breakthrough innovations using examples from different sectors.

References

Barsoux, J.-L., Bouquet, C., & Wade, M. (2022). Why outside perspectives are critical for innovation breakthroughs. *MIT Sloan Management Review.* https://sloanreview.mit.edu/article/why-outside-perspectives-are-critical-for-innovation-breakthroughs/.

IBM. (2021). The new AI innovation equation. https://www.ibm.com/watson/advantage-reports/future-of-artificial-intelligence/ai-innovation-equation.html.

Luo, J. (2022). Data-driven innovation: What is it? https://arxiv.org/ftp/arxiv/papers/2201/2201.08184.pdf.

Paschen, U., Pitt, C., & Kietzmann, J. (2020). Artificial intelligence: Building blocks and innovation typology. *Business Horizons*, 63(2), 147–155.

Sun, B. & Liu, Y. (2021). Business model designs, big data analytics capabilities and new product development performance: Evidence from China. *European Journal of Innovation Management*, 24(4), 1162–1183.

Index

Printed in the United States
by Baker & Taylor Publisher Services